W9-CZT-701

ELECTRONIC SWITCHING CIRCUITS

Boolean Algebra and Mapping

PRENTICE-HALL SERIES IN ELECTRONIC TECHNOLOGY

Dr. Irving L. Kosow, editor

Charles M. Thomson, Joseph J. Gershon. and Joseph A. Labok, consulting editors

MATTHEW MANDL

Temple University

ELECTRONIC SWITCHING CIRCUITS

Boolean Algebra and Mapping

Prentice-Hall, Inc., Englewood Cliffs, N.J.

Current printing (last digit):

10 9 8 7 6 5 4 3

13-252163-6
Library of Congress Catalog Card No.: 76-84844
Printed in the United States of America

PRENTICE-HALL INTERNATIONAL, INC., *London*
PRENTICE-HALL OF AUSTRALIA, PTY. LTD., *Sydney*
PRENTICE-HALL OF CANADA LTD., *Toronto*
PRENTICE-HALL OF INDIA PRIVATE LTD., *New Delhi*
PRENTICE-HALL OF JAPAN, INC., *Tokyo*

PREFACE

This text discusses the principles of electronic switching circuits for combinational as well as sequential systems. It emphasizes the electronic aspects related to the design of computer logic circuitry, industrial control switching, and other allied areas of usage.

All the basic methods for switching syntheses are discussed in detail and illustrated. Numerous examples are included to help the reader understand the methodology. Questions and problems at the end of each chapter provide additional exercises for the reader, and allow him to evaluate how well he has assimilated the material. Combinational switching is covered in the first eight chapters, and the sequential switching methods (which require the combinational theory as a prerequisite) are covered in Chapters 9 and 10.

Initial chapters introduce the fundamentals of switching logic, positive and negative factors, logic maps and Venn diagrams. Solid-state circuitry is analyzed with emphasis on the *logical switching factors* so these principles can be applied to modules, chips, or integrated circuitry without involvement in design variations of any particular individual units.

Multilevel switching is covered in Chapter 3, and Boolean-algebra principles, theorems, postulates, and laws are explained in Chapter 4. Chapter 5 relates to binary numbers, codes, and switching processes as applied to adders.

Chapter 6 details the processing and simplification procedures of the Boolean-algebra expressions obtained from truth tables. Both the minterm and maxterm aspects are discussed, and a number of practical-aspect problems are included.

Additional simplification procedures are covered in Chapter 7, where the use of multivariable Karnaugh maps is explained and illustrated. Numerous practical examples serve to clarify complex map-simplification methods.

Flip-flop circuits, parallel adders, and storage devices are introduced in Chapter 8 and serve as a foundation for the analysis of sequential switching systems wherein these devices are employed in conjunction with the combinational switching covered in earlier chapters.

Chapter 9 introduces the sequential systems, describes the symbols and terms used, and covers the initial synthesis procedures which include the verbal problem statement, the preparation of the appropriate flow diagram and table for both the pulse and level-type systems. Again, a number of practical examples help the reader acquire facility in these procedures.

Chapter 10 covers elimination of redundant expressions in maps and tables, procedures for handling secondary assignments, and factors relating to outputs. The derivation of the final expression is explained for both the Boolean-algebra and mapping methods. The preparation of the appropriate switching-system diagrams representative of the finalized switching design is detailed, illustrated by drawings and representative practical examples.

MATTHEW MANDL

Yardley, Pennsylvania

CONTENTS

4.

BOOLEAN ALGEBRA PRINCIPLES

5.

NUMBERS, ADDERS, AND CODES

6.

PROCESSING BOOLEAN EXPRESSIONS (COMBINATORIAL CIRCUITS)

7.

SIMPLIFICATION WITH MULTIVARIABLE MAPS 121

8.

FLIP-FLOPS, PARALLEL ADDERS, AND STORAGE 153

9.

SEQUENTIAL CIRCUITS (INTRODUCTION) 170

10.

SEQUENTIAL CIRCUITS (SYNTHESIS) 186

CONTENTS

1

CIRCUIT LOGIC

INTRODUCTION

Digital computers, modern telephone systems, and industrial electronic control, all utilize complex switching circuits for performing their tasks in an economical and efficient manner. When one considers a single switch of any of the foregoing systems, it appears as a simple device, having either an "off" or an "on" state. If several switches are connected in series or parallel, the functional characteristics of the network so formed will still be readily apparent from inspection and no serious problems are evident in the design of an operational system to meet requirements.

The switching methods in computers and telephone systems, however, are not always based on simple "do" or "don't" situations, but are often actuated by *logic* demands, where an "on" condition will result only if certain other conditions prevail. This factor, plus the necessity for routing numerous signals into many multipath channels, means that the organization and design of the complicated end-result *cannot* be readily achieved by educated guessing, or assembled by intuition alone.

While competent designers may be able to produce a functional system by such procedures, much time may be wasted and the final circuitry often contains more switches than necessary to perform the required functions. Thus, the system is less reliable, more costly to assemble and maintain, and less efficient because of the unnecessary additional circuits.

1

Because the switching must be in accordance with numerous logical conditions, the design and assembly of such systems require an orderly approach involving certain rules of logic. A most useful tool for expediting the formation of efficient switching systems and for analyzing their functional capabilities, is *Boolean algebra,* named after its originator, George Boole, the famous English logician and mathematician (1815–1864). This logic algebra is concerned with the manipulation of variables which are limited to only two values: *one* and *zero.* These values relate to the *on* and *off* of an electric or electronic switch.

Supplementing Boolean Algebra are also various logic diagrams and maps which provide a graphic representation of specific switching states and also are of considerable value in the production of effective switching systems.

In this and subsequent chapters, the basic circuit logic is presented first, followed by Boolean expressions and equivalent map representations. Finally their usage in logic circuit analysis, simplification, and design, are covered in detail.

ELECTRONIC SWITCHING FACTORS

There are three basic switch types:

1. The manually operated type, such as the pushbutton or toggle
2. The electric relay, operated by current flow through a solenoid
3. The vacuum-tube or solid-state electronic circuit, triggered by steady-state or pulse signals.

It is the solid-state electronic switch in which we are primarily interested. Such a switch can be in wired form, or manufactured as a chip, module, or other integrated unit. Such switches are far superior to the mechanical type because of their extremely rapid switching cycles, their dependence on electronic characteristics rather than mechanical, and their space-saving compactness.

The "on" and "off" characteristics of a simple switch are related to logic by designating the "on" condition as a "true" state and the "off" as a "false" state. The logical Boolean statements of true and false are more conveniently notated as 1 and 0 in switching systems, and thus conform to the binary system of notation described later. Boolean algebra variables can also be identified by letter representations, such as *A, B, C, D,* etc. Thus, if an electronic switching function is represented by a variable designated as *A*, then $A = 1$, if the statement is true. If the statement is false, then $A = 0$.

Thus, our basic switch can have the following logic representations:

On	Off
True	False
Yes	No
A	0
1	0

Logic 1 and Logic 0

In switching systems it is important that the relationships of logic 1 or logic 0 to signal-voltage polarities be understood. Either a positive signal or a negative signal can be assigned a logic 1 value (but not both in the same system). If the design engineer selects a positive voltage to represent logic 1; then zero voltage (or a negative voltage) would represent logic 0. These positive and negative signal voltages are *relative to each other* and not necessarily positive or negative with respect to circuit ground. Thus, with a positive voltage representative of logic 1, a *less positive* voltage represents logic 0, since it is in the negative direction, and hence negative with respect to the logic 1 signal.

Similarly, if a negative voltage is selected to represent logic 1, then a voltage positive with respect to the negative signal would represent logic 0; as would a zero voltage, since the latter is positive with respect to the logic 1 signal voltage.

When a positive signal voltage represents logic 1, it is sometimes referred to as the *up-state* signal and the negative signal voltage for logic 1 is called the *down-state*. Other terms used for these states are *up level* and *down level*.

These factors are illustrated in the two circuits shown in Fig. 1-1. As shown in (A), a PNP transistor is used, with the input signal applied between the base and emitter, and the output is obtained from across the 5100-ohm resistor between emitter and ground. The collector is bypassed with a capacitor (0.1 μF) to prevent signal-voltage variations in the collector circuit. Such a circuit is known as an *emitter follower* and compares to the vacuum-tube *cathode-follower* circuit.

3

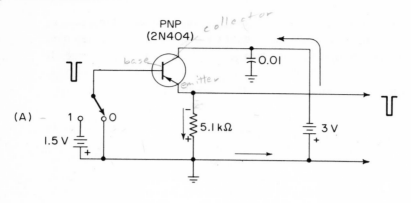

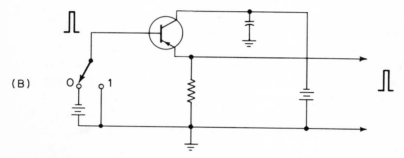

Fig. 1-1. Emitter-follower logic.

In switching systems, the purpose of the emitter follower is to provide for isolation between certain other circuits as well as maintaining proper impedance match between the output of one logic circuit and the input of another. The term *follower* stems from the fact that the phase of the output signal "follows" that of the input signal. The term *grounded collector* has also been applied to this circuit and it is of particular value in all branches of electronics.

The emitter-follower characteristics include low output impedance, wide frequency response, and sufficient signal current output to drive several other logic circuits.

For the circuits shown in Fig. 1-1, conduction prevails if the base is biased in a forward direction with respect to the emitter, that is, the base must be negative with respect to emitter polarity. The collector, on the other hand, must be reverse biased, so that the collector P zone is negative with respect to the emitter. With most transistors the

4

absence of forward bias between base and emitter, or the application of reverse bias (base positive), will cause the transistor to cut off and prevent current flow between collector and emitter.

For the circuits shown, NPN transistors could also be used, though battery polarities would be opposite to those shown, as would the logic 1 and 0 representations.

If we elect to use a negative signal voltage to represent logic 1, the circuit arrangement as shown in (A) would be appropriate. Here, steady-state voltages are used for simplicity in initial basic analysis. A logic 0 input condition exists when the input switch arm is to the right as shown. If this switch is now momentarily tripped to the left, a negative voltage is applied between base and emitter and forward bias is thus established. Now conduction between emitter and collector rises to a high level and electrons flow·in the direction shown by the arrows. In consequence, a voltage drop occurs across the emitter resistor with a polarity as shown, producing an output signal representative of logic 1. If we had designated our input signal as A; then $A = 1$.

Suppose, however, that we initially have the switch in the logic 1 position and momentarily trip it to the right (logic 0 position in Fig. 1-1A). This applies a positive signal to the input of the emitter follower, and the resultant reverse bias will reduce conduction between the emitter and collector and the voltage drop across the emitter resistor will decline. Thus the output *change* is in a positive direction, producing a positive-polarity output signal. Since we have selected a negative signal as our logic 1, the positive-polarity output signal represents logic 0.

If, however, we had initially selected a positive signal polarity to represent logic 1, then the circuit arrangement shown in (B) would be used. Now, the switch position at the left establishes our negative voltage as logic 0, and a momentary tripping of the switch to the right to engage the logic 1 terminal results in the application of reverse bias to the input. Hence the decline in voltage across the emitter resistor which produces a positive voltage change is representative of logic 1.

INVERTER (NOT) CIRCUIT

In the emitter-follower circuit we found that the phase of the output signal "follows" that of the input, and no phase inversion occurs between the output signal and that applied to the input. In the conventional grounded-emitter amplifier, however, the phase of the output signal is *not* the same as that of the input signal, and hence the logic expression for such a circuit is NOT *circuit*. A typical amplifier is shown in Fig. 1-2A, using an NPN transistor. As shown, the emitter

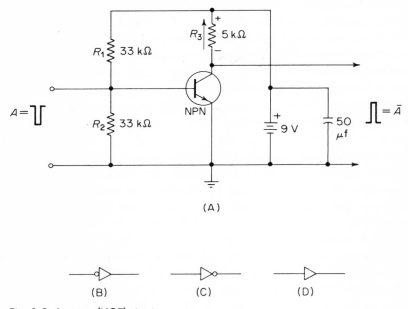

(A)

(B)   (C)   (D)

Fig. 1-2. Inverter (NOT) circuit.

is negative with respect to the collector because the battery applies its negative potential to the emitter via the common ground lead. The positive potential is applied to the collector, with R_3 as the output load resistor. The necessary reverse bias for the collector is thus satisfied.

The forward bias for the base-emitter circuit is furnished by the two resistors, R_1 and R_2, which act as voltage dividers across the battery. Since both resistors have equal values, a positive 4.5-V potential appears at the base of the transistor, making it positive with respect to the emitter.

If a negative-polarity pulse is applied to the input as shown, it has the effect of reducing the forward bias and thus decreasing transistor conduction between collector and emitter. In consequence, the voltage drop across the load resistor R_3 drops, and the collector voltage rises toward the positive battery potential. Thus, an amplified positive-polarity output signal is produced as shown.

6

For a positive-polarity input pulse the phase-inverting characteristic would still prevail, since the positive signal applied to the input would increase the forward bias and increase conduction. The result is a larger voltage drop across resistor R_3, and a drop of potential at the collector. Thus, a negative-polarity output signal is obtained.

The NOT function is also referred to as *complementing*, because a signal of opposite characteristics (the complement) is produced. In Boolean terms, this is a negation or falsehood type expression. If we designate the input signal as A; then the output signal is *not A*. The complement or negation of a statement is indicated by a line over the symbol (an overbar). Thus, $\overline{A}$ means *not A* and, if A represents 1, the expression $\overline{A}$ indicates 0. Similarly $\overline{1} = 0$, and $\overline{0} = 1$. If $A = 0$, then $\overline{A} = 1$.

In logic switching algebra only two truth values are used (true and false, or 1 and 0, etc.). Hence, a double negation such as *not (not A)* is logically equal to A. A double negation sign has been used on occasion to signify this: $\overline{\overline{A}} = A$. A prime sign has also been used with a symbol to denote the NOT function, such as A'. In this text, however, the line over the symbol (overbar) will be used.

Subsequent chapters will relate the logic NOT function to other Boolean expressions and algebraic manipulations of switching circuits.

Standard symbols for the NOT circuit are shown at Fig. 1-2B and C, and either one can be used, as preferred. The symbol shown in (D) indicates an amplifier without the inverting function.

BASIC LOGIC SWITCHES

Before considering logic electronic switches, we will analyze some of their electric counterparts for a foundation in understanding the logical concepts which apply. Consider, for instance, the parallel switching illustrated at Fig. 1-3A. Here is a basic doorbell circuit, where a front door switch (A) *or* a back door switch (B) will cause the chimes (C) to ring. Logically, we can then state that either A *or* B rings the chimes, *or* both. Hence, the logic switching-circuit terminology for such a circuit is an OR circuit or an OR gate, since *A or B = C*.

Even though we are using three different symbols here, it must be remembered that we have only two states to consider, *on* and *off*. Hence, we would state that on = 1 and off = 0, and 1 *or* 1 = 1, indi-

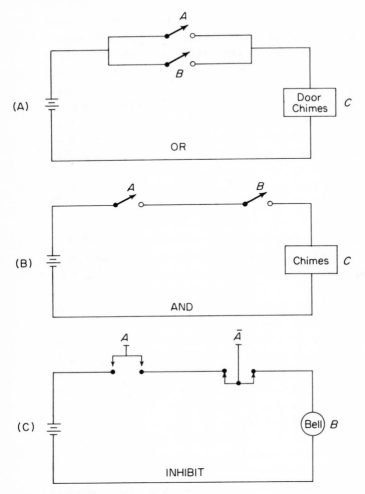

Fig. 1-3. Basic logic circuit functions.

cating that if either *or* both switches are in the *on* (1) state, the chimes will also be in the *on* or 1 state.

In Boolean algebra, the + sign is used as a *logical connective* to denote the OR function. Thus, $A + B$ replaces the "*A or B*" statement. The + sign, however, *does not indicate arithmetical addition* but simply represents the OR function. Other symbols have also been used to indicate the OR logic, including ∪ as well as ∨. However, here we will use the + sign for OR.

Figure 1-3B shows another doorbell-type system, where A could be the front door switch, and B a master switch to disengage the system when the occupants are not home. Here, both the A switch *and* the B switch must be closed to ring the chimes. Logically, we can then state that A *and* $B = C$. Hence, the logic switching-circuit terminology for such an arrangement is an AND circuit or AND gate. Because both A and B must be actuated to close the circuit, this arrangement has also been called a *coincidence* gate, because of the necessity for coinciding closure of both A and B switches. Using logic 0 and 1 we then have: 1 *and* 1 = 1.

The logical connective for expressing the AND function is the multiplication sign, though as with the $+$ sign for the OR function, the sign is not taken in the literal arithmetic sense. Hence, the multiplication sign should be considered as *and* when we read $A \cdot B = C$, or $1 \cdot 1 = 1$. As in standard algebraic notation, the multiplication sign can be omitted and the logical connective implied by placing the letters close together as $AB = C$. The symbols $\cap$ and $\wedge$ have also been used for the logical AND function, but herein we will use the multiplication sign.

In (C) a different sort of doorbell arrangement is shown. Here, switch marked $\overline{A}$ is a pushbutton *normally closed* type. If the normally open switch A is depressed, the bell will ring. If switch $\overline{A}$ is depressed, however, it opens the circuit and the bell can't ring, even though switch A is depressed at the same time as switch $\overline{A}$. Thus, switch $\overline{A}$ has an *inhibiting* function where A *and* (*not* A) = 0.

Using our logical connective for the AND function, we can express this logic as $A \cdot \overline{A} = 0$. Substituting logic 1 and 0, we have $1 \cdot \overline{1} = 0$. Note that $\overline{A}$ alone = 0, because it opens the circuit. If $\overline{A}$ only is used, we can substitute 0 for A and get: $\overline{A} \cdot 0 = 0$. If we use A alone [A but *not* (*not* A)], the expression is: $A \cdot \overline{\overline{A}} = B$, or $1 \cdot \overline{\overline{1}} = 1$.

This preliminary association of the NOT function with other logic expressions indicates its important modifying aspects in Boolean algebra. An understanding of the basic logical principles outlined so far will aid materially in assimilating the more complex Boolean manipulations which are covered in later chapters.

The following logical statements summarize the factors covered in this chapter and indicate usage of the logical connectives and the overbar.

AB	means	A AND B
$A \cdot B$	means	A AND B
$A + B$	means	A OR B
$A\bar{B}$	means	A AND NOT B
$A \cdot \bar{B}$	means	A AND NOT B
$A + \bar{B}$	means	A OR NOT B
$\bar{A} \cdot B$	means	B AND NOT A
$(\bar{A}B) + C$	means	$(B$ AND NOT $A)$ OR C
$A + (B \cdot \bar{C})$	means	A OR $(B$ AND NOT $C)$
$A + (\bar{B}C)$	means	A OR $(C$ AND NOT $B)$

SYMBOLS

Symbols for the amplifier and the NOT circuit were shown in Fig. 1-2. The symbols for the AND circuit, the OR circuit, and the INHIBITING circuit are shown in Fig. 1-4. These logic circuits are often designed

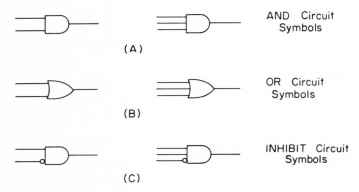

(A)   AND Circuit Symbols

(B)   OR Circuit Symbols

(C)   INHIBIT Circuit Symbols

Fig. 1-4. Logic-circuit symbols.

for three inputs (or more) and are represented by additional input lines as shown.

Figure 1-4A shows the symbols for the AND circuit (coincidence switches) which will be used throughout this text. Other symbols having a somewhat different body shape have been used, but the ones shown have found wide acceptance.

Two- and three-input OR circuits are shown in Fig. 1-4B. These are readily distinguished from the AND circuits because of the curved line at the input terminals, and the less rounded body at the output.

When the inhibit function is to be indicated, and AND-circuit symbol is used, as shown in (C). The inhibit input terminates at a small circle at the main body of the symbol, as illustrated. Again, a two-input circuit may be employed, or three or more inputs can be utilized as required to obtain the necessary logical functions.

Questions and Problems

1. How are the *on* and *off* states of a switch related to logic statements of true and false?

2. Describe what is meant by the terms *up-state signal* and *down-state signal*.

3. What is the purpose of an emitter follower?

4. What are the general characteristics of an emitter follower?

5. Draw an emitter-follower circuit of the type shown in Fig. 1-1, using an NPN transistor with a positive-polarity signal representing the logic 1 state.

6. Explain briefly why a conventional grounded-emitter amplifier circuit is known as a NOT circuit in logic switching terminology.

7. Draw a grounded-emitter amplifier of the type shown in Fig. 1-2, using a PNP transistor. Show direction of electron flow in the emitter-collector circuit.

8. a. If $A = 0$, what is the value of $\bar{A}$?
   b. How do we indicate double negation for a symbol such as A?

9. Briefly explain the characteristics of a logic OR circuit.

10. Briefly explain the characteristics of a logic AND circuit.

11. Explain what is meant by logical connectives and illustrate their usage with alphabetical symbols.

12. Explain the characteristics of an INHIBIT circuit.

13. Explain the logical meaning of the following:

   $\bar{A} + B$; $A + (BC)$; $(A + B)C$; $AB\bar{C}$; $(A + B)(\bar{C} + D)$.

11

14. Draw symbols for the following expressions:

$$A + B = C; \quad ABC = D; \quad A\bar{B} = C; \quad A = A.$$

15. How does a NOT symbol differ from the symbol for an amplifier? Include drawings of the symbols in your explanation.

16. How does the NOT symbol resemble the INHIBIT symbol? Is there a relationship between the NOT function and the INHIBIT function?

2

SOLID-STATE LOGIC SWITCHES

INTRODUCTION

The AND circuits and the OR circuits introduced in Chap. 1 are of the solid-state variety when used as logic switching in electronic systems. Solid-state components permit much more rapid switching, are devoid of the faults inherent in mechanical devices, and can be manufactured in subminiature form as chips or modules. Diodes form switching circuits containing a minimum of parts, and hence are less costly than transistors. The advantage of transistors is the signal amplification obtained, plus the fact that the complement feature is available (logic negation).

The AND, OR, and NOT circuits are analyzed and described by representing their characteristics in Boolean expressions, truth tables, Venn diagrams, and Karnaugh maps. When such circuits are combined to perform specific logic processes, algebraic manipulations permit circuit simplification.

The important aspects of solid-state logic switches are covered in this chapter. The related discussions will serve as a foundation for the study of Boolean algebra principles and applications covered in subsequent chapters.

POSITIVE AND NEGATIVE LOGIC

In Chap. 1 the logic 1 and logic 0 were described, with reference to up-state and down-state factors. When a logic 1 is represented by positive signals, the term *positive logic* is applied to the operation of the circuit. Diode switching using positive logic is shown in Fig. 2-1 for the OR and AND gates. To simplify the discussion, steady-state voltages are shown. Functionally, these operate in the same manner as those illustrated for pulse waveforms discussed later.

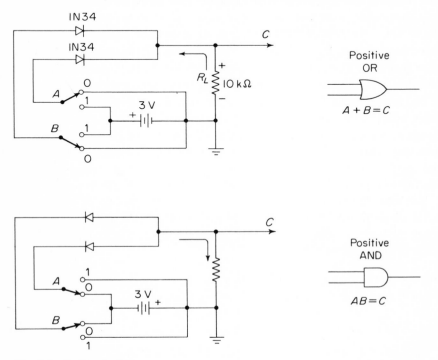

Fig. 2-1. Positive logic.

For the OR circuit shown in Fig. 2-1, electron flow is up through the load resistor (R_L) as shown by the arrow. Current will flow when the battery is in the circuit to supply the necessary forward bias for the diodes. Switches A and B are shown connected to the "0" terminals; thus applying ground to both sides of the diodes. This is the logic "0" state and neither diode conducts. If switch A is now tripped to the logic 1 position, the necessary forward bias is applied to the lower diode and conduction occurs. The voltage drop across the load resistor rises, producing a positive output signal voltage.

Similarly, if the B switch is set to the logic 1 position, the positive voltage applied to the left of the upper diode will cause conduction also

SOLID-STATE LOGIC SWITCHES

and produce an output signal. Obviously, if both switches are placed in the logic 1 position, both diodes conduct and an output signal is again obtained. Thus, this is an OR circuit having the characteristics $A + B = C$.

This OR circuit (as well as the three-input types discussed later in this chapter) are known as INCLUSIVE OR switches because the *true* statements occur not only for A alone, *or B* alone, but also for both. Thus, the INCLUSIVE OR logic is *A or B or both $A + B$*. If three inputs A, B, C, are involved, the logic for the inclusive OR is: *A or B or C, or* all inputs applied simultaneously. These factors will be more clearly understood by a study of the truth tables which follow.

The term INCLUSIVE OR is used to distinguish this switch from the EXCLUSIVE OR type in which the logic is *A or B, but not both inputs at the same time.* The EXCLUSIVE OR circuit is also known as a *half adder*, since it performs binary addition without carry and the logic is $A + B = 0$. Because the EXCLUSIVE OR circuit is a multilevel type, it is discussed in Chap. 3, *Multilevel Logic Switching.*

The logic functions of switching circuits can be expressed in a *truth table* which shows the various combinations of input signals and the resultant logic output. For the OR circuit shown in Fig. 2-1 a truth table can be constructed as follows:

$$0 + 0 = 0$$
$$A + 0 = C$$
$$0 + B = C$$
$$A + B = C$$

Instead of A and B, we can show the truth table with logic 1's and 0's:

$$(A + B) = C$$

$$0 + 0 = 0$$
$$1 + 0 = 1$$
$$0 + 1 = 1$$
$$1 + 1 = 1$$

If the diodes and the battery polarity are reversed as shown on the lower switching circuit in Fig. 2-1, a positive-logic AND circuit is formed. Now with the A and B switches in the logic 0 position, the necessary forward bias is applied to the diodes and conduction occurs. Electron flow is in the direction shown by the arrow, raising the output

15

voltage to a negative level above ground because of the voltage drop across the load resistor. As long as either diode conducts, current will flow and a steady-state output voltage prevails.

If the A switch is set to the logic 1 position, the lower diode will stop conducting. Since, however, the upper diode still has the forward bias applied, it conducts and holds the output voltage level above ground. Similarly, if B switch alone is applied to the logic 1 terminal, the upper diode stops conducting, but the lower diode still conducts and produces an output voltage above ground. If, however, *both* switches are set to logic 1, both diodes stop conducting and the output voltage at C drops to the ground (positive) level, thus producing a positive output voltage. Consequently, we procure an output only if both A *and* B are in the positive "1" position, thus forming a two-input AND switch. The truth tables that now apply are as follows:

$$0 \cdot 0 = 0$$
$$A \cdot 0 = 0$$
$$0 \cdot B = 0$$
$$A \cdot B = C$$

Using logic 1's and 0's for the truth table, we obtain:

$$(A \cdot B) = C$$

0 0	=	0
1 0	=	0
0 1	=	0
1 1	=	1

Thus, only when coincidence occurs for input signals is an output signal obtained for the positive AND switch indicated.

In Fig. 2-2 the use of negative logic is illustrated. Note that the AND circuit is identical to the OR circuit shown in Fig. 2-1, and the negative logic OR circuit is also identical to the AND circuit of Fig. 2-1. Only the logic differs. For the negative logic AND switch of Fig. 2-2, the positive battery potentials are assigned the logic 0, and when both the A and B inputs are connected to these terminals both diodes obtain the necessary forward bias for conduction. In consequence the C output is above ground level and is positive because of the voltage drop across the load resistor.

16

If either the *A* or *B* switch is set to the logic 1 position, one of the diodes will stop conducting, but because the other diode still conducts, the *C* output is still above ground. When both *A and B* are set to logic 1 input, both the anode and cathode of each diode are at ground potential and neither diode conducts. In this instance the *C* output drops to the negative ground level producing a negative output. Thus, when a *negative* signal coincidence occurs at the input, a *negative*

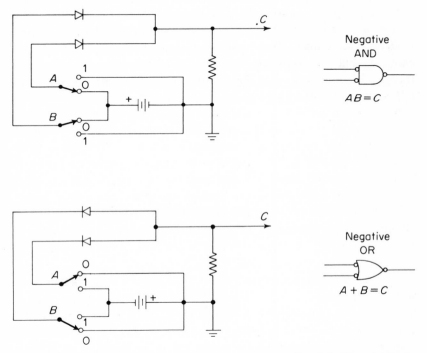

Fig. 2-2. Negative logic.

output signal is produced. The truth table for the negative-logic AND circuit (using 1's and 0's) is the same as that shown for the positive-logic AND circuit. Sometimes, however, the negative-logic symbol shown at the right of the circuit is used to indicate the negative-logic AND switch. If negative signals are used for *A* and *B*, then *A* and *B* still represent 1's as in positive logic. Positive and negative logic must not be confused with negation. The symbol $\overline{A}$ means *not A* and hence 0, regardless of whether negative or positive logic prevails.

17

The negative-logic OR circuit is shown in the lower drawing of Fig. 2-2. Here, the *A* and *B* inputs are initially set at the ground level (0), and neither diode conducts. If the *A* input is set to the logic 1 position it applies the necessary negative potential to the lower diode and conduction occurs. The voltage across the output load resistor rises and the *C* output rises above the ground level to a negative potential. Similarly, a *B* input alone will produce an output, as will *A or B*. Thus, the application of one or more negative inputs will produce a negative output for the negative-logic OR function. The symbol for the negative-logic OR circuit is shown at the right.

Diode-circuit logic (DCL) can also be used with three inputs, as shown in Fig. 2-3. Here, the duality of circuit function is again illustrated, with either circuit of Fig. 2-3 able to function as an AND or an OR switch. For the upper circuit, a negative-logic AND gate is formed if negative input pulses are employed for logic 1. If a negative pulse is applied to the *A* input alone, it will overcome the positive potential supplied by the battery and the uppermost diode stops conducting. Since the other two diodes are still conducting, the output voltage at *D* is still positive and above ground. Similarly, no output signal change is procured for a *B* input, or a *C* input alone. When, however, negative input signals are applied simultaneously to the *A*, *B*, and *C*, terminals, all diodes stop conducting, and the output voltage drops to the ground level, producing a negative output signal. Thus, a negative-logic AND circuit function is obtained.

If a positive input signal is considered as logic 1 and applied at *A*, it will increase conduction for the upper diode, provided the input signal has a greater positive amplitude than exists normally between input and ground. With increased conduction, the voltage drop across the output resister rises to a higher positive value, producing a positive logic 1 output pulse. Similarly, a positive pulse applied to *B or C*, or at all inputs at once, produces a positive output signal at *D*, resulting in the positive-logic OR function. (For the OR function the battery potential can be omitted, and the three resistors at the input placed at ground at their common junction.)

For the lower circuit shown in Fig. 2-3, the diodes and battery potential are reversed, forming either a negative-logic OR circuit, or a positive-logic AND circuit, as shown. A negative signal at one or more inputs increases conduction and produces a negative output signal at the output terminal *D*, forming a negative-logic OR circuit. If, however, positive input signals are used, coincidence must prevail

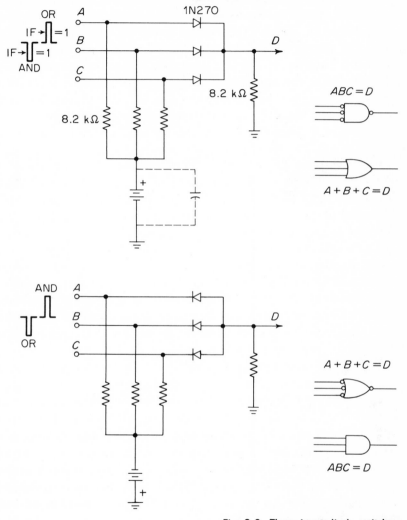

Fig. 2-3. Three-input diode switches.

at all inputs to prevent any diode from conducting. When conduction stops, the output voltage at terminal *D* drops to ground level (positive) and a positive-logic AND circuit is formed.

19

For a three-input OR circuit, the truth table appears as follows:

$$A + B + C = D$$

A	B	C	D
0	0	0	0
1	0	0	1
0	1	0	1
0	0	1	1
1	1	0	1
0	1	1	1
1	0	1	1
1	1	1	1

Note that an output is obtained for any single signal or combination of input signals. For the AND circuit, however, an output is produced only if all inputs are present, as shown by the truth table:

$$A \cdot B \cdot C = D$$

A	B	C	D
0	0	0	0
1	0	0	0
0	1	0	0
0	0	1	0
1	1	0	0
0	1	1	0
1	0	1	0
1	1	1	1

LOGIC DIAGRAMS AND MAPS

Diagrams are also employed to illustrate truth logic graphically. These are composed of shaded or unshaded squares or circles to represent specific logic 1 and logic 0 conditions. Such truth diagrams are also called *logic maps* or *logic charts*. Map diagrams to express logic statements are not new. Charles L. Dodgson (1832–1898), the English mathematician and author, described some in his books *The Games of Logic* and *Symbolic Logic*. (Dodgson's pen name was Lewis Carroll and under this pseudonym he wrote *Alice in Wonderland*, etc.) The logic maps are sometimes called *Karnaugh maps* after Maurice Karnaugh who in 1953 published *The Map Method for Synthesis of Combinational Logic Circuits* (*Transactions*, AIEE, C & E, Vol. 72,

November 1953). Another contribution to map and chart logic representations had been made by Edward W. Veitch who published *A Chart Method for Simplifying Truth Functions* (*Proceedings, Association of Computing Machinery*, 1952).

Logic map representations for some basic circuit symbols are shown in Fig. 2-4. As shown in the upper left diagram, horizontal rows

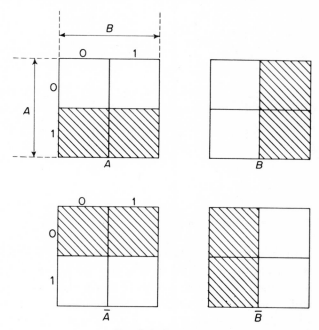

Fig. 2-4. Symbol designations for maps.

represent logic 1 or logic 0 inputs for *A*. Vertical rows represent 1 and 0 inputs for *B*. If zero areas are shaded, as shown in the lower drawings, they represent negation, as $\overline{A}$ for the lower left, and $\overline{B}$ for the lower right.

Maps for typical basic circuit logic are illustrated in Fig. 2-5. The following rules apply:

1. The intersection of the horizontal and vertical rows represents the logic statement for the selected input logic.
2. The intersecting square is shaded if true, and blank if false.
3. When the AND function is involved, any square not coincident is left blank.
4. When the OR function is involved, all shaded squares remain intact.

21

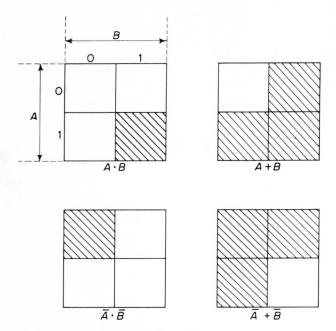

Fig. 2-5. Logic map applications.

In the function $A \cdot B$ shown at the upper left in Fig. 2-5, if the vertical 1 is selected for B and the horizontal 1 for A, the intersection is at the lower right of this map. Because this is shaded, the statement is true and represents the AND-circuit function $1 \cdot 1 = 1$. Note that the 1 areas that are not coincident are left blank, according to rule 3 above.

The OR function $(A + B)$ is shown at the upper right map. Here again the horizontal row represents the logic 1 for A and the vertical row the logic 1 for B. According to rule 4, however, all shaded squares remain, since either A or B, or both $A + B$ will provide a logic 1 output.

The lower left map shows the $\overline{A} \cdot \overline{B}$ function. Since two negations are involved, the square diagonally opposite to the shaded square in the $A \cdot B$ map shows the logic coincidence which prevails. The $\overline{A} + \overline{B}$ map is shown at the lower right, and here the unshaded area is diagonally opposite that which prevails for the map of the $A + B$ function.

Additional examples are given at the upper section of Fig. 2-6. The $A \cdot \overline{B}$ function is illustrated in the upper left map. Since the 1 function for A is represented by the lower two horizontal areas, and the negation for B as the left vertical squares, the intersection is at the

22

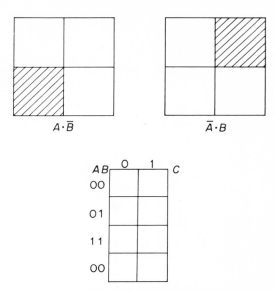

Fig. 2-6. Two- and three-variable maps.

lower left as shown, with other areas left blank as determined by rule 3. If $\overline{A} \cdot B$ is involved, the map is as shown at the upper right. The negation for A occupies the upper horizontal squares, and the 1 function for B the right vertical squares, producing the shaded area indicated.

These two-input maps (and the Venn diagrams which follow) help visualize and illustrate the logic involved in the true and false OR and AND functions. For logic-circuit simplification procedures, the three-input map is more useful. This is shown in preliminary form in Fig. 2-6 for comparison purposes. The formation of such logic maps is more complex and will be taken up in detail later in Chap. 7.

Logic diagrams can also be constructed using circles to express the true or false functions for 1 and 0, as shown in Fig. 2-7. These are called *Venn* diagrams after the 19th-century mathematican John Venn. The $A + B$ function is shown at the upper left. Overlapping shaded circles are used for the OR function, with shaded areas representing both the 1 states and the intersecting "true" area. The $A \cdot B$ function is shown at the upper right, with only the intersecting area shaded to indicate the AND state, according to the same rule 3 used for the square-area maps. The AND function for ABC is shown at the lower left, with only the intersection of the three circles shaded to indicate that a true (1) condition alone prevails for coincidence of all three

23

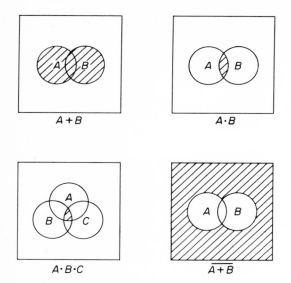

Fig. 2-7. Venn diagrams.

values of *A*, *B*, and *C*. The $\overline{A + B}$ function is shown at the lower right. Here, the surrounding area is shaded to indicate the negation. This type negation for AND and OR functions will be discussed more fully later and illustrated with De Morgan's theorem.

The Venn diagrams are included herein for reference only as examples of another graphical method for illustrating logic functions. Starting with Chap. 4, the rectangular logic maps and standard symbol diagrams will be used to analyze logic circuit functions and demonstrate circuit simplification.

TRANSISTOR LOGIC CIRCUITS (TLC)

While diode logic circuits are inexpensive and simpler in design, transistors offer the advantages of signal gain and high-speed switching characteristics. The emitter follower discussed in Chap. 1 is used to form OR and AND logic as shown in Fig. 2-8. Again, positive or negative logic can be used, as required.

For the upper circuit shown in Fig. 2-8, NPN transistors are used, though PNP types could also be employed with a reversal of battery potential and logic.

24

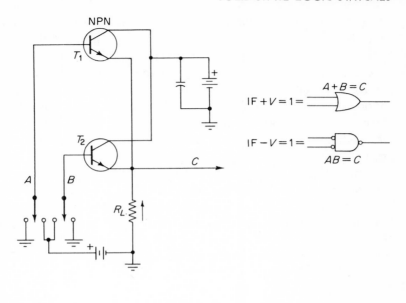

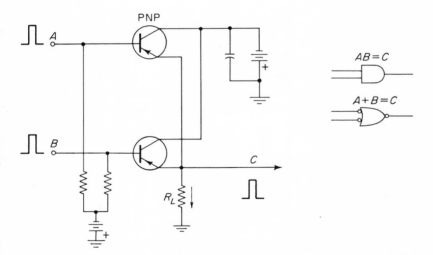

Fig. 2-8. Transistorized OR and AND logic.

If the positive terminal of the battery is used for logic 1 and ground for logic 0, an OR switch is formed. If either (or both) *A* or *B* are flipped to the logic-1 state, the necessary forward bias is applied between the base-emitter circuits and conduction occurs, with electron flow

25

direction as shown by the arrow beside the load resistor R_L. Consequently, a voltage drop occurs across the load resistor, and the output at C rises in a positive direction above ground.

If a positive input represents logic 0 and a negative input logic 1, an AND circuit is formed. With both the A and B switches at the positive terminals, both transistors conduct and a steady-state potential exists at the C terminal output (positive above ground). Now both A *and* B must be placed in the ground (negative) position to cause a signal-change output. When coincidence occurs the output voltage at terminal C drops to the ground level, producing a negative-polarity output signal. Thus, a negative-logic AND circuit function prevails.

The principle illustrated in Fig. 2-8 is important because it indicates the dual function that can be obtained from a single circuit by reversing the logic (from positive to negative). If a positive (P) signal represents logic 1, and a negative (N) signal represents logic 0, a truth table for the OR function appears as:

OR switch

$$N + N = N$$
$$P + N = P$$
$$N + P = P$$
$$P + P = P$$

For the same circuit, however, the use of a negative (N) signal for logic 1, and a positive (P) signal for logic 0, produces an AND circuit and the truth table would be:

AND switch

$$P \cdot P = P$$
$$N \cdot P = P$$
$$P \cdot N = P$$
$$N \cdot N = N$$

For the OR switch where a positive (P) signal is logic 1, a P-signal output will be obtained for either one or both inputs of a positive signal. For the AND switch, however, a negative (N) signal is logic 1, hence we must have logic-1 input coincidence for a negative-signal output, or $N \cdot N = N$. (*Note:* both truth tables express logic duality; only the logical connectives differ.) Each table contains a P output for one or more positive-signal inputs, as well as an N output for coincidental negative-signal inputs. (The subject of duality is covered in greater detail in Chap. 4, Laws of Dualization.)

The lower circuit of Fig. 2-8 is a practical version of the transistorized gate, suitable for pulse-signal input. For positive logic, an AND circuit is formed for the PNP transistor circuit. Both transistors are supplied the required forward bias for the emitter-base circuits and reverse bias for the emitter-collector circuits as shown. Thus, both transistors conduct and cause a steady-state negative voltage to appear across the common emitter resistor R_L.

If a positive pulse signal is applied to the A input, the forward bias for the upper transistor is nullified and conduction for this transistor stops. Conduction, however, still occurs in the lower transistor. When pulses are applied to both the A and B inputs, both transistors stop conducting and the voltage drop across the output resistor R_L declines to the ground level (positive). Thus, a positive change occurs across R_L to produce a logic-1 output for coincidence at the input.

If the logic is reversed by using negative signals for logic 1, an OR circuit is formed. If a negative-polarity signal is applied to either (or both) of the inputs, forward bias is increased and conduction increases. Greater conduction through R_L raises the voltage drop and produces an output-voltage change in the negative direction for a logic-1 output signal. Again, a comparison of the two truth tables will indicate the duality of the circuit when the logic is reversed:

AND switch	OR switch
$P \cdot P = P$	$N + N = N$
$P \cdot N = N$	$N + P = N$
$N \cdot P = N$	$P + N = N$
$N \cdot N = N$	$P + P = P$

NOR AND NAND CIRCUITS

The phase-inverting characteristics of the grounded-emitter circuit were illustrated in Fig. 1-2 and the related discussions pointed out that this is a logic NOT circuit (negation) where the phase of the output signal is *not* the same as the phase of the input signal. Consequently, when such a circuit is used to form logic switches, the NOT function is combined with the OR logic or the AND logic.

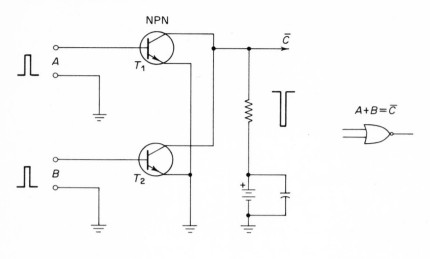

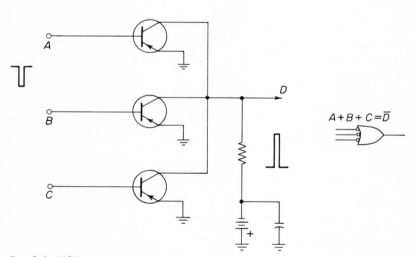

Fig. 2-9. NOR circuits.

If the grounded-emitter circuit is used to form OR logic as shown in the upper diagram of Fig. 2-9, the output signal will be inverted. In the absence of an input signal each NPN transistor is cut off. For an input at either *A or B*, the necessary forward bias is applied and conduction occurs, producing a voltage drop across the output resistor which develops a negative-polarity signal as shown.

28

(If steady-state forward-bias voltages were applied to these transistors, negative logic could be applied to form an AND circuit, as discussed earlier. To help clarify the discussions of the NOR and NAND circuits, however, only one logic version of each circuit will be analyzed.)

Because the collectors are connected in parallel, either A, B, or both $A + B$ produce conduction. Since the output signal is inverted, however, a NOT function also prevails, forming an OR NOT circuit. A common term for such a circuit is a NOR circuit (NOT OR). The logic, therefore, is: $A + B = \bar{C}$, with a symbol designation as shown to the right of the drawing.

The lower drawing shows the use of PNP transistors with negative logic. Here, negative input signals are used to apply the necessary forward bias for the base-emitter circuits. Because of the inverted (NOT function) output, however, a NOR circuit is still formed. The symbol, however, is altered to indicate the negative-logic NOR function as shown at the right.

By placing the collector of one transistor in series with the emitter of the other, as shown in Fig. 2-10, AND logic is formed. Since the

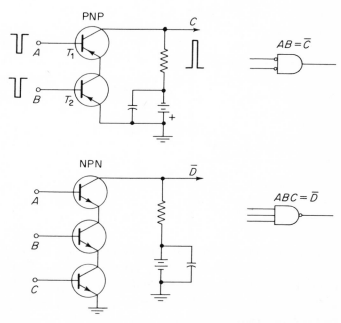

Fig. 2-10. NAND circuits.

29

output is inverted, however, an AND NOT condition exists, and this circuit is called a NAND circuit (NOT AND).

For the upper PNP circuit of Fig. 2-10, the application of a negative signal at the T_1 transistor (A) input will not cause conduction because the emitter is in series with the collector of transistor T_2 which is not conducting. Similarly, if a negative signal is applied to the B input, transistor T_2 has the necessary forward bias applied, but the conduction path through T_1 is blocked because the latter does not conduct. Only when signals are applied to both A and B inputs will conduction occur. Because negative logic is used for this NAND circuit the symbol has small circles at the inputs, as shown at the right. The logic expression is $AB = \overline{C}$.

A three-input NAND circuit is shown in the lower drawing of Fig. 2-10. Positive logic is used with the NPN transistors, and input signal coincidence produces an output signal. When all transistors conduct, the resultant current-flow through the output load resistor produces a negative-signal output to form the logic $ABC = \overline{D}$. The symbol representation is shown at the right.

In a NOR circuit the logic expression $A + B = \overline{C}$ states, in essence, that either A or B produces a negation, hence it follows that the logic expression $\overline{A + B} = C$ is also valid. Since a NOR function is involved. a negated input will be inverted and will produce a C output. Similarly, the NAND function of $AB = \overline{C}$ has a counterpart in $\overline{AB} = C$. Such relationships are evident from inspection of the NAND and NOR truth tables, where the logic 1's represent the A, B, and C values, and 0's the negations, or $\overline{A}$, $\overline{B}$, and $\overline{C}$ values.

NAND $A \cdot B = \overline{C}$			NOR $A + B = \overline{C}$		
0	0	1	0	0	1
0	1	1	0	1	0
1	0	1	1	0	0
1	1	0	1	1	0

Note that the first logical expression in the NAND truth table is $0 \cdot 0 = 1$. Using negated symbols for 0's we have: $\overline{AB} = C$. For the second expression $0 \cdot 1 = 1$ we obtain $\overline{A}B = C$. The third expression $1 \cdot 0 = 1$ yields $A\overline{B} = C$ and the last expression $1 \cdot 1 = 0$ gives us $AB = \overline{C}$. Thus, this truth table is identical to that which would be

30

written for the expression $\overline{AB} = C$. Similar factors apply to the NOR truth table.

The NAND expression $\overline{AB} = C$ implies that the logical connective is also negated ($\overline{A \cdot B} = C$) which would convert the AND function ($\cdot$) to the OR (+) and hence $\overline{AB} = C$ is equivalent to the NOR $\overline{A} + \overline{B} = C$. Similarly, $\overline{A + B} = C$ is equivalent to $\overline{A} \cdot \overline{B} = C$. These factors are the basis for an important rule in switching logic (De Morgan's rule), which is covered in greater detail in Chap. 4.

MULTIPLE-INPUT REPRESENTATIONS

From an inspection of the diode and transistor switching circuitry covered in this chapter, it is evident that additional components can be added to the units shown to increase the number of inputs. Thus, multiple-input OR, AND, NOR, and NAND circuits can be formed as required to meet specific design demands. Some representative types are illustrated in Fig. 2-11 in symbol form, with the logic that applies indicated for each. All these can be considered as *single-level*

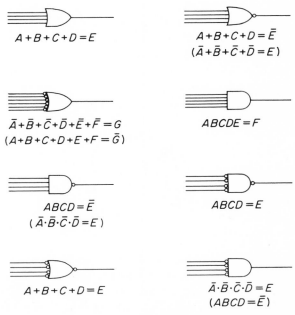

$$A + B + C + D = E$$

$$A + B + C + D = \overline{E}$$
$$(\overline{A} + \overline{B} + \overline{C} + \overline{D} = E)$$

$$\overline{A} + \overline{B} + \overline{C} + \overline{D} + \overline{E} + \overline{F} = G$$
$$(A + B + C + D + E + F = \overline{G})$$

$$ABCDE = F$$

$$ABCD = \overline{E}$$
$$(\overline{A} \cdot \overline{B} \cdot \overline{C} \cdot \overline{D} = E)$$

$$ABCD = E$$

$$A + B + C + D = E$$

$$\overline{A} \cdot \overline{B} \cdot \overline{C} \cdot \overline{D} = E$$
$$(ABCD = \overline{E})$$

Fig. 2-11. Multiple-input logic.

31

logic, since all inputs are applied to a single logic circuit. When AND and OR circuits feed other logic circuits, *two-level* logic is utilized, and such combinational forms are discussed in Chap. 3.

The positive-logic and negative-logic symbols are useful to indicate whether positive signals indicate logic 1's or negative signals represent the 1's. The logic expression, however, is the same, as mentioned earlier. For a four-input NAND circuit, the expression $ABCD = \overline{E}$ does not indicate either positive or negative logic, only the particular logic applicable to the NAND circuit. The same holds true for $\overline{ABCD} = E$. The symbol alone indicates which polarity is utilized to represent logic 1 and which polarity logic 0.

Also, a NOT circuit has no effect on the polarity of the *logic* employed. If a positive logic is used, a positive voltage always represents logic 1 and the NOT circuit will produce a negative signal, but this represents NOT 1 ($\overline{1}$), which is actually logic 0.

Truth tables for the multi-input switches follow the same pattern as for the two-input types given earlier. A representative table is given below for reference. (Rows could be in binary sequence if desired.)

$$A + B + C + D = E$$

A	B	C	D	E
0	0	0	0	0
0	0	0	1	1
0	0	1	0	1
0	0	1	1	1
0	1	0	0	1
0	1	1	1	1
0	1	0	1	1
1	0	0	0	1
1	0	0	1	1
1	1	0	0	1
1	0	1	1	1
1	1	1	0	1
1	1	0	1	1
1	0	1	0	1
0	1	1	0	1
1	1	1	1	1

For multiple-input AND switches, coincidence inputs must be present to obtain an output. Thus, $1 \cdot 1 \cdot 1 \cdot 1 = 1$, but $1 \cdot 0 \cdot 1 \cdot 1 = 0$.

RTL AND DTL LOGIC SWITCHING

By combining resistors with transistors we can also obtain multiple-input logic functions as shown in the upper drawing of Fig. 2-12. Here,

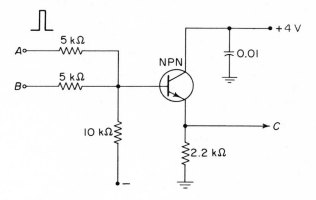

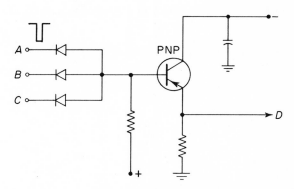

Fig. 2-12. RTL and DTI switching.

the resistor-transistor logic (RTL) forms an OR circuit. An NPN transistor is used, and for conduction the base emitter must be forward biased, with the base positive with respect to the emitter. Note, however, that a negative potential is applied to the base. This reverse bias will not permit the transistor to conduct. When a positive-polarity

33

signal is applied to either input *A*, or *B*, or both, it overcomes the negative potential at the base and the transistor conducts. The conduction (with electrons flowing from ground to the emitter and collector) produces a voltage drop across the output resistor which is positive at the emitter. Since this is also the output terminal, a positive-polarity output signal is developed. Additional resistors could, of course, be used at the input circuit for a greater number of inputs.

The diode-transistor logic (DTL) switch shown in the lower drawing of Fig. 2-12 uses a PNP transistor. Except for polarities and negative logic, the operational characteristics are similar to the RTL circuit. With a dc-positive potential applied to the base the transistor is cut off. When one or more input signals are applied, the negative potential (logic 1) overcomes the positive voltage at the base and the necessary forward bias is developed to permit conduction. Since electron flow is from collector to emitter and down through the output resistor, a negative output signal is produced, providing a logic-1 output.

The RTL and DTL circuits shown offer the advantages of transistor switching, circuit isolation, and emitter-follower impedance matching as discussed in Chap. 1. If grounded-emitter type circuits are used, the amplified signal output will have a NOT function producing NAND or NOR switches as the ones discussed earlier.

A DTL circuit is shown in Fig. 2-13 which illustrates the INHIBIT

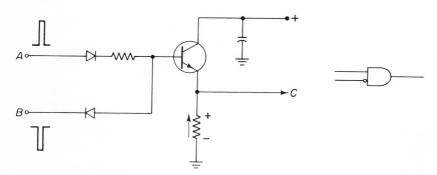

Fig. 2-13. The INHIBIT (*but-not*) logic.

(*but-not*) logic function. Here, an NPN transistor is used but no forward bias is present. Thus, in the absence of an input signal at *A* the transistor does not conduct. When a positive signal (logic 1) is applied at input *A* as shown, it furnishes the necessary forward bias (positive base with respect to the emitter) and conduction occurs, producing a positive-polarity output signal. If a negative signal is applied at input B alone, the negative polarity increases the reverse bias and the transistor is driven more into the nonconduction region and hence no output signal is obtained. (Since a negative-polarity signal here represents logic 0, the input is designated as $\bar{B}$.) Similarly, if both A and $\bar{B}$ are applied, no conduction occurs because the negative-polarity $\bar{B}$ input cancels the A input (assuming the input signals have the same relative amplitude with respect to their polarities). Thus $\bar{B}$ *inhibits* any input at A and the logic is A *but not* B $(\bar{B})$: $A = C$, $A\bar{B} = 0$, $\bar{B} = 0$.

Logical Expressions

Heretofore, a single variable has been shown as the output of a logic circuit. Often, however, the output of a logic gate is shown in multiple-variable form to indicate more clearly the logic functions which are performed by the switching. Thus, if variables A and B are applied

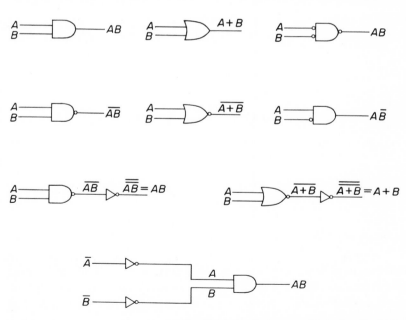

Fig. 2-14. Logical expressions—input-output.

35

to an AND switch, the output would be indicated as AB which shows the true "*A and B*" function. Similarly, the output from an OR switch would be designated as $A + B$, while the output from a NAND switch would be shown as $\overline{AB}$ or as $\overline{A \cdot B}$, as in Fig. 2-14. For the inhibiting function, the output is expressed as $A\overline{B}$, indicating the "*A but not B*" logic.

As also shown in Fig. 2-14, the NAND function can be followed by a NOT circuit to convert the expression to its original form. The NAND switch produces $\overline{AB}$ which is applied to the NOT circuit to produce a double negation $\overline{\overline{AB}}$, which is equivalent to the original expression AB. Similarly, the NOR function can be followed by a NOT circuit to revert the expression $\overline{A + B}$ to its original form, and thus maintain the respective positive or negative logic which prevails throughout a system.

Two NOT circuits can also be used to invert variables to their appropriate logic polarity as shown in the lower drawing of Fig. 2-14. Here, the input variables are negated by the NOT circuits to change $\overline{A}$ to A and $\overline{B}$ to B as shown. Thus, the AND-switch output is AB.

Questions and Problems

1. Briefly explain the difference between positive logic and negative logic.

2. How can an AND switch be used to produce OR logic?

3. What symbol is used to indicate negative logic for the expression $A + B + C = D$?

4. Show a logic map formed by squares for the function $\overline{A} + \overline{B}$.

5. Show a Venn diagram for the logic function $A + B + C$.

6. Show a Venn diagram for the logic function $\overline{A + B + C}$.

7. What advantages do transistorized logic switches have over diode types?

8. Explain what is meant by a NOR circuit and a NAND circuit.

9. Draw the symbols for a three-input NOR circuit using positive-

logic input, and a three-input NAND circuit using negative-logic input.

10. Draw a schematic of a three-input NOR circuit using NPN transistors. How can this be made to function as a NAND circuit?

11. Show the symbol for the nand logic $\overline{A} \cdot \overline{B} \cdot \overline{C} \cdot \overline{D} = E$, and prepare a truth table for this logic function.

12. Which of the following expressions represent NOR or NAND functions:

$$ABCD = E; \quad \overline{ABCD} = \overline{E}; \quad A + B + C + D = \overline{E};$$
$$\overline{A} + \overline{B} + \overline{C} + \overline{D} = E; \quad \overline{A} \cdot \overline{B} \cdot \overline{C} \cdot \overline{D} = E;$$
$$\overline{A} + \overline{B} + \overline{C} = D; \quad A \cdot B \cdot C = \overline{D}.$$

13. Show the symbol for the expression $\overline{A} + \overline{B} + \overline{C} = D$ for negative logic.

14. Show the symbol for the expression $\overline{A} + \overline{B} + \overline{C} = D$ for positive logic.

15. Explain what is meant by RTL and DTL switching circuits.

16. Draw an RTL circuit with three inputs, with the transistor functioning as a NOT circuit.

17. What is meant by a *but not* function?

18. Is the inhibit circuit related to the OR switch or to the AND switch? Explain on what factor you base your answer.

19. Assume that three NOT circuits precede a three-input OR switch. If the input variables are $\overline{A}$, $\overline{B}$, and $\overline{C}$, what is the output expression from the OR switch?

20. If the output from a NOT circuit is $\overline{A + B + C}$, what is the equivalent expression?

3

MULTILEVEL LOGIC SWITCHING

In the logic circuitry involving telephone switching, industrial control, telemetering, and digital computers, combinations of logic AND, OR, and NOT circuits are used to perform more complex functions than obtained from individual switches. Thus, one or more AND circuits may feed OR circuits, or several OR circuits may have their output lines connected to the inputs of AND gates, etc. Often combinations of OR and AND circuits may feed other logic circuits, which in turn have their outputs applied to still other logic switches. Thus, two-level or multilevel circuitry is involved and the logic functions as well as the Boolean expressions become more complex than heretofore shown.

By understanding the logic that applies to such multilevel circuits and by becoming familiar with the Boolean statements that apply to them, a foundation is laid for the study of specific principles that are used to reduce and simplify complex logic circuitry. Thus, this chapter introduces the higher level switching systems and the logic statements that apply, as well as the basic Boolean algebra covered in Chap. 4.

TWO-LEVEL DLC

A two-level diode-logic circuit with three inputs is shown in Fig. 3-1. Here diodes D_1 and D_2 form OR inputs for $A + B$. Diodes D_3 and D_4

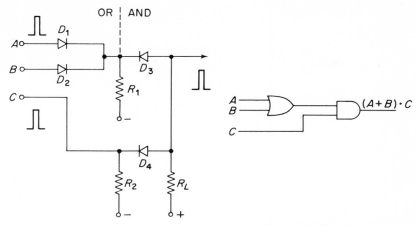

Fig. 3-1. Two-level DLC with three inputs.

are reversed in comparison with the first two diodes, and hence form AND circuits for the positive logic applied to the inputs. With a negative potential applied to the bottom of R_1, and a positive voltage at R_L diode D_3 has the necessary forward bias and conducts, holding the output at a steady-state negative level above the positive ground. Diode D_4 also has the necessary forward bias applied for conduction.

If a positive-polarity signal pulse is applied at the A input, diode D_1 conducts and the voltage drop across R_1 increases. Thus, the voltages at both the cathode and anode of D_3 approach the ground (positive) level and the diode stops conducting. Because diode D_4 still conducts, however, the output is held at the negative level above ground. If a positive pulse is applied to the C input only, diode D_4 will stop conducting but has little effect on the output voltage level.

When the C input is combined with A or B, or both A as well as B, both diodes D_3 and D_4 stop conducting and a voltage drop no longer appears across the output resistor R_L. Consequently the output level rises to the positive ground level. Thus, an output is obtained only for $(A + B) \cdot C$, as shown by the logic-symbol diagram at the right in Fig. 3-1.

As for the single-level logic circuits discussed in Chap. 2, we can use the same basic circuit and reverse the logic to obtain an AND function instead of an OR, and an OR function instead of an AND, as shown in Fig. 3-2. Here negative logic is employed and the input becomes an AND switch, while the output diodes D_3 and D_4 are now

39

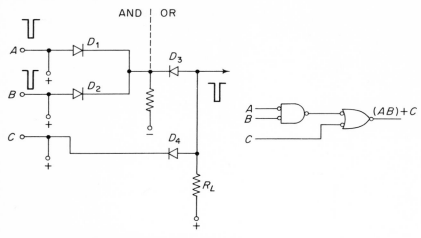

Fig. 3-2. Two-level DLC with negative logic.

an OR switch. Steady-state positive voltages are applied at the inputs, and in the absence of an input pulse signal, diodes D_1 and D_2 conduct, causing the cathode and anode potentials for D_3 to be sufficiently close to each other to prevent conduction. For diode D_4 the positive input potential also holds this diode at nonconduction.

Only if negative signal voltages are applied simultaneously to the A and B inputs will diode D_3 conduct. When this occurs, the current flow through R_1 causes a voltage drop which produces a negative-polarity output signal. A negative-polarity pulse signal at C alone will produce an output because it overcomes the positive potential and permits D_4 conduction. Again the voltage drop that results across R_L produces an output signal. Thus our logic is: $(AB) + C$. The negative-logic symbol representation for this circuit is shown at the right in Fig. 3-2.

Two-level DLC Variations

By reversing the diodes and polarities for the circuit shown in Fig. 3-1, we obtain the negative logic equivalent as shown at the upper left of Fig. 3-3. The inputs A, B, C, are now negative-polarity pulses. Because these now represent logic 1, there is no change in the logic statement $(A + B) \cdot C$. As pointed out earlier, negative logic does not indicate *negation*, and the overbar ($\overline{A}$) is used only for the NOT function.

40

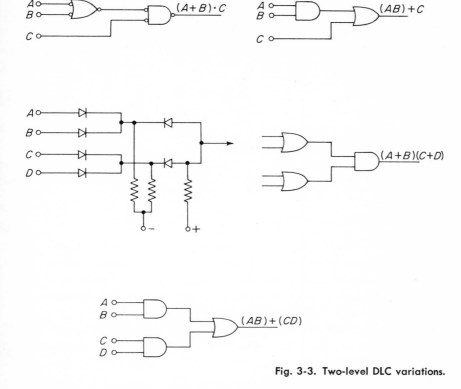

Fig. 3-3. Two-level DLC variations.

Similarly, the diodes and polarities can be reversed in Fig. 3-2 to produce a positive-logic equivalent switch, as shown in symbol form at the upper right in Fig. 3-3. Again the same logic expression prevails but positive-polarity signals are involved.

By duplicating the input diodes in either of the circuits shown in Fig. 3-1 and Fig. 3-2, a four-input two-level circuit can be formed, as shown at the center left of Fig. 3-3. Here two OR switches feed a single AND switch, with circuit function essentially identical to that shown in Fig. 3-1. Now the logic expression must involve *A or B and C or D*: $(A + B)(C + D)$.

If the *C* input of Fig. 3-2 is also replaced by a two-input diode circuit, the four-input resultant would perform the logic $(AB) + (CD)$. For a positive-logic circuit of this type the symbolic diagram is as shown in the lower drawing for Fig. 3-3.

The single inputs *C* of Fig. 3-1 and Fig. 3-2 can be placed at the

41

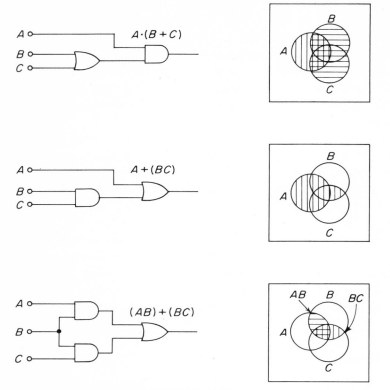

Fig. 3-4. Two-level logic and Venn diagrams.

top for the A input with a change of logic as shown in Fig. 3-4. For the upper diagram the logic expression is $A \cdot (B + C)$ as shown, where the input A must now be present with either B *or* C to produce an output. The Venn diagram shown at the right will help to understand the logic function obtained. The circle for A contains vertical shading lines, while the circles for B and C contain horizontal shading lines. Where the three circles intersect, small squares are formed to indicate the true condition. Thus, A *and* B *or* C is the logic expression illustrated. Note that the intersection of circles A and B forms a true statement, as does the intersection of A and C. For the intersection of the B and C circles, however, no squares are formed and the statement $B + C$ only is false.

For the center logic diagram the expression $A + (BC)$ is obtained and the Venn diagram for this logic is also shown at the right. As

42

shown earlier in Fig. 2-7, the OR-function circles can be distinguished from the AND functions by representing the latter as unshaded. This is done for the center Venn diagram in Fig. 3-4. The shaded circle represents the A logic, and the unshaded B and C circles the AND logic. Intersection shading represents a true condition. Thus the intersection of B and C is a true statement, A alone is a true statement, and the complete logic is $A + (BC)$, because the intersection of all three shows a shaded area also.

Often two or more inputs of logic circuitry are combined into a single input with a single expression. This is shown in the lower drawing of Fig. 3-4, where one of each of the AND circuits is connected to the other to form a common input designated as B. The logic thus produced is $(AB) + (BC)$, because A and B or B and C will produce an output. The appropriate Venn diagram is shown at the right, with blank circles representing the AND functions of AB or BC, and shaded areas representing true conditions. Note that the intersection of circles B and C provide a shaded (true) area, as do the intersection of circles A and B. Also, where all three intersect, the shaded area represents the logic $(AB) + (BC)$.

MULTIPLE-INPUT TWO-LEVEL LOGIC

As many inputs as necessary can be used with the two-level DLC circuits. A five-input circuit is shown in Fig. 3-5 and consists of an upper three-input OR switch *and* a two-input lower OR switch all of which combine to form the expression $(A + B + C)(D + E)$. The representative symbolic diagram is shown at the right.

The circuit arrangement need not necessarily be as shown in the upper schematic of Fig. 3-5. Often a *matrix* arrangement is used as shown in the lower drawing. This circuit is identical to the one shown in the upper drawing. Only the representation differs, with the components arranged in symmetrical fashion, horizontally and vertically in matrix form.

In addition to variations in circuit layout, the symbol representations may also be arranged differently on occasion. One method is shown in Fig. 3-6 which is referred to as a *logic tree*. Multiple inputs and logic gates are displayed horizontally along the top, and as less and less successive logic gates are fed, they are placed beneath as shown

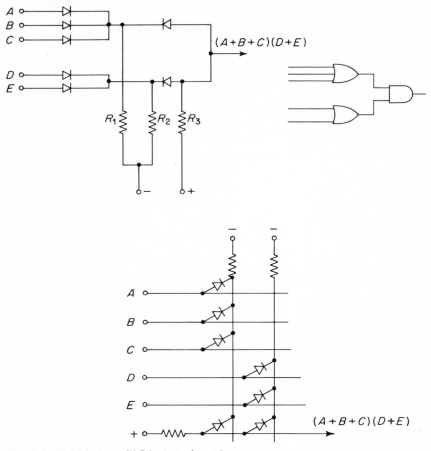

Fig. 3-5. Multiple-input DLC logic and matrix.

in a formation resembling a tree. All inputs for the diagram shown are to AND switches which, in turn, feed subsequent OR circuits to perform the logic $(AB) + (CD) + (EF) + (GH)$. Here, an eight-input three-level diagram is represented.

Two-level TLC

In Fig. 3-7 transistor logic is employed for two-level operation with NOR and NAND switching. The input NOR circuit is the same as

44

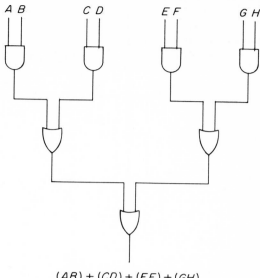

$$(AB) + (CD) + (EF) + (GH)$$

Fig. 3-6. Logic tree.

illustrated earlier in Fig. 2-9, which with reverse logic becomes a NAND circuit as shown at the output of Fig. 3-7.

In the absence of forward bias at the input, transistors T_1, T_2, and T_3 are in the nonconducting state. For the NAND-circuit transistors T_4 and T_5, the necessary forward bias is applied through the base resistors, making the base terminals negative with respect to the positive (ground potential) emitters. During conduction of T_4 and T_5 the voltage drop across the output resistor in the collector circuits causes the output line to be near ground potential (positive).

When an input signal (negative potential for logic 1) is applied to either the A or B inputs, or both, conduction occurs for the T_1 or T_2 (or both) transistors and the voltage drop across the collector resistor rises, producing a positive output signal. This, in turn, is applied to the input of the T_4 transistor, and applies a reverse bias which cuts off T_4. Since, however, T_5 still conducts, there is no change in the output.

If a negative-polarity signal is applied at the C input alone, transistor T_3 will conduct and apply a positive potential at the base of T_5, causing this transistor to stop conducting. Transistor T_4, however, still conducts and no output signal is produced. When we have the $A + B$ input

45

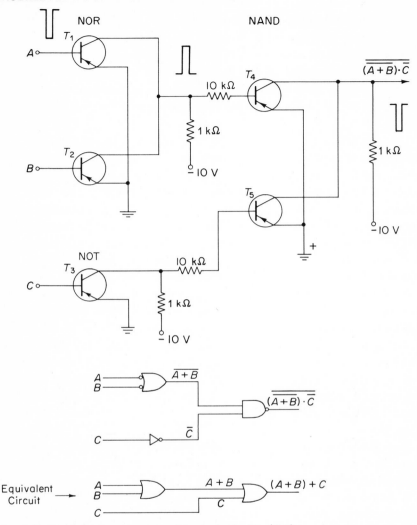

Fig. 3-7. Two-level three-input transistor logic circuitry (TLC).

and the C input, both T_4 and T_5 are cut off and current no longer flows through the output load resistor. Consequently the output voltage drops to the -10-V level producing a negative-polarity output signal as shown.

The symbolic representation for this circuit is also shown in Fig. 3-7. Because the C-input transistor T_3 inverts the phase of the signal it is a NOT circuit and the appropriate symbol for this is included in the

46

symbolic representation. Also, since the $A + B$ inputs are to a NOR circuit, phase inversion also occurs. Thus, the input to transistor T_4 is $\overline{A + B}$ and the input to T_5 is $\overline{C}$. Because the output circuit is a NAND switch, reversal of phase occurs again and we obtain the expression $\overline{(A + B) \cdot C}$. Note the use of the negative-logic symbol for the NOR switch, and the positive-logic symbol for the NAND switch.

The expression $\overline{(A + B) \cdot C}$ is the true logical statement for the circuit shown in Fig. 3-7 and a reduction of the double negations would result in an expression representative of *equivalent* logical circuitry. This comes about because the NOR, NAND, and NOT circuits invert the *entire* expression, *including the logical connectives*. Thus, $\overline{A + B}$ differs from the expression $\overline{A} + \overline{B}$ because the negation of the logical connective in $\overline{A + B}$ forms the equivalent expression $\overline{A} \cdot \overline{B}$. Similarly, $\overline{A \cdot B}$ differs from $\overline{A} \cdot \overline{B}$, because the negation of the logical connective in $\overline{A \cdot B}$ makes the expression equivalent to $\overline{A} + \overline{B}$.

As mentioned in Chap. 1, a double negation of a variable reverts it to its original form, so that $\overline{\overline{A}} = A$, $\overline{\overline{1}} = 1$, etc. Thus, eliminating the double negation of $A + B$ and C in the expression $\overline{\overline{(A + B)} \cdot \overline{C}}$ and changing the logical connective AND to OR, produces $(A + B) + C$ which is representative of the equivalent circuit shown at the bottom of Fig. 3-7. Here, the $A + B$ inputs are applied to one OR circuit and the output in combination with the C input, applied to a second OR circuit, producing $(A + B) + C$. This circuit could be reduced to a single three-input OR switch.

Logical equivalence, duality, and allied factors are covered in greater depth in Chap. 4, where the specific laws, theorem, and postulates of Boolean algebra are discussed.

A four-input transistor-logic circuit arrangement is shown in Fig. 3-8. The input transistors have no forward bias applied and hence are cut off. To cause conduction the input must be a positive-polarity signal (positive logic). If positive logic is to be used for the output transistors (T_7 and T_8) also, phase inversion (NOT) circuits must be used between the NOR and NAND gates as shown, consisting of T_5 and T_6.

Forward bias is applied to the NOT circuit transistors and these conduct. The voltage drops across the output resistors R_3 and R_4 of the NOT circuits cause the base circuits of T_7 and T_8 to be near the ground potential (negative). Depending on the type transistors and

47

MULTILEVEL LOGIC SWITCHING

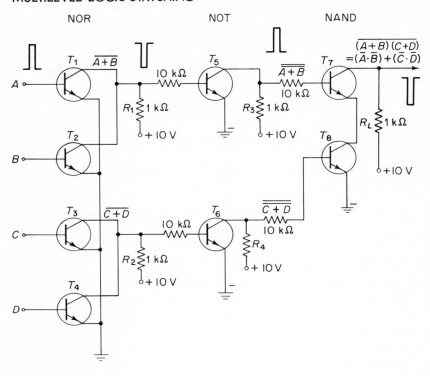

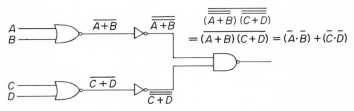

Fig. 3-8. Four-input TLC.

circuit parameters used, the output transistors may conduct at a low level or be cut off entirely. Since the collector of T_8 is in series with the emitter of T_7, current flow through the output load resistor R_1 can occur only if both output transistors conduct. (See Fig. 2-10 and related discussions.) With both T_7 and T_8 cut off, no voltage drop occurs across R_L and the steady-state output voltage is at a positive level.

If a positive pulse is applied to the A input, the NOR function of transistor T_1 produces a negative-signal output. The negative voltage appearing at the base of T_5 applies reverse bias to this NPN transistor

48

and cuts it off. The voltage at the base of T_7 rises (positive direction) and causes this transistor to be in a conduction possibility. Because transistor T_8 does not conduct, however, it acts as an open circuit for the current and no output signal is produced.

When A or B inputs are present at the same time that C or D inputs are applied, both bases of output transistors T_7 and T_8 have increased positive potentials and hence conduction occurs. The voltage drop across R_L decreases the positive-output potential toward the ground (negative) level, producing a negative-polarity signal as shown. The symbolic representation is also shown in Fig. 3-8.

Two-level DTL

Diodes can be combined with transistors in two-level circuitry to obtain additional signal amplification and for performing NOT or other logic functions. A typical arrangement is shown in Fig. 3-9, where a

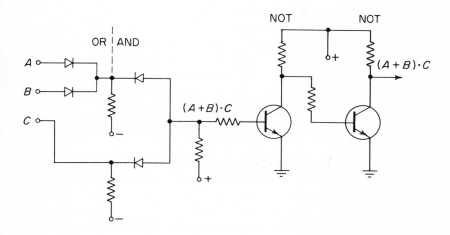

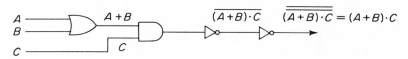

Fig. 3-9. Two-level DTL.

DTL OR AND system is followed by two transistor NOT circuits. The OR and AND circuits function as described earlier for Fig. 3-1. Positive logic is used and the output from the AND circuit applies a positive-polarity signal to the input of the first transistorized NOT circuit. The signal is inverted and a negative-polarity signal is applied to the output transistor. This decreases conduction and the output voltage rises in a positive-polarity direction. Thus, the output is positive logic, as was the case with the input, but signal amplification has occurred. Thus the logic for this circuit is $(A + B) \cdot C$, and would be the same without the NOT circuit amplifiers.

Multilevel Logic

It will be understood by now that the symbolic representations of logic functions do not indicate whether diode logic circuitry (DLC), transistor logic circuitry (TLC), or combinations of these are used. Since the basic logic is the same for DTL, DLC, or TLC, only the logic function given by the symbol is necessary. Often the relative polarity of the logic may be ignored where its indication is not necessary, and the positive-logic symbol for the logic circuit may be shown, even though negative signals are used. Where the negative logic has an effect on the overall logic expression it must, of course, be represented by the appropriate negative-logic symbol. The NOT function, since it indicates negation, must be included to keep the logical sequences in perspective.

Two multilevel examples are shown in Fig. 3-10 to summarize the logical sequences that may be encountered. For the upper illustration, five inputs are used, with $AB + C$ applied to the upper circuit section and DE to the lower NAND circuit. Thus, the application of positive signals to the $AB + C$ inputs represent logic 1's, while the logic 1's for the DE inputs are negative-polarity signals. A negative-logic NOR is present at the output. So far as the logic developments and progressions are concerned in this circuit, the negative-logic NAND and NOR switches could have been represented by positive-logic symbols.

The $(AB) + C$ expression at the output of the OR circuit is inverted by the NOT circuit to $\overline{(AB) + C}$, and the NAND switch produces $\overline{DE}$, as shown. When these expressions are applied to the NOR circuit

50

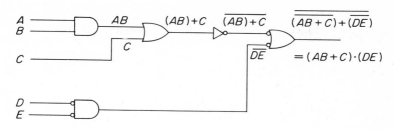

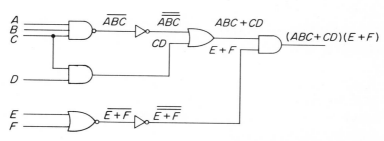

Fig. 3-10. Multilevel logic examples.

an output expression $\overline{\overline{(AB + C)} + \overline{(DE)}}$ is produced, which is equivalent to $(AB + C)\cdot(DE)$.

A multi-input, multilevel system having a common input to the two AND circuits is shown in the lower drawing. Here, six inputs are present out of a possible seven because of the C input being common to both AND switches. Note, however, that a C input for the upper logic circuit undergoes a NOT function because of the NAND circuit, while a C applied to the lower AND gate produces a logic 1 output. Thus, when ABC coincidence occurs for the NAND circuit the output is $\overline{ABC}$ as shown. For the CD input to the AND circuit, a CD output is applied to the OR circuit as shown. The NOT circuit preceding the OR gate transforms the negated $\overline{ABC}$ to $\overline{\overline{ABC}}$, and because a double negation reverts the expression to its original form of ABC, the OR circuit output is $ABC + CD$.

The $E + F$ inputs to the NOR circuit produce an output of $\overline{E + F}$ which is inverted by the NOT circuit to the double negation $\overline{\overline{E + F}}$, which is equivalent to $E + F$. Hence, the output AND switch receives $(ABC + CD)\,and\,(E + F)$ for the logic output of $(ABC + CD)(E + F)$.

51

EXCLUSIVE OR

So far the OR switches have been of the *inclusive* OR type, where *A or B or C or* all inputs at one time provide a true statement. As mentioned in Chap. 2, there is another OR switch which is of the EXCLUSIVE OR type wherein *A or B but not* both, produce a true statement. The EXCLUSIVE OR switch is a two-level type as shown in Fig. 3-11.

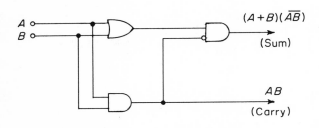

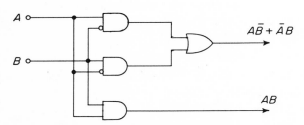

Fig. 3-11. Exclusive OR (half adder) circuits.

For the upper EXCLUSIVE OR switch shown in Fig. 3-11, OR, AND, and INHIBITOR circuits are used. If only an *A* input is applied, an output is obtained from the OR circuit and fed to the INHIBITOR input. In the absence of an inhibiting signal, an output is obtained. Since there was no signal coincidence at the AND circuit, no output was obtained in the carry line. Similarly, if only a *B* input is applied, an output is also obtained satisfying the *A + B* logic.

If both the *A* and *B* signals are applied, an output is obtained from both the OR circuit and the AND circuit. Consequently the INHIBITOR circuit has a coinciding input and no output is developed. At the same time the carry line furnishes an output *AB* as shown. Thus, the

EXCLUSIVE OR circuit will provide an output for either *A or B*, *but not* for both. The logic expression is, therefore, $(A + B)(\overline{AB})$. Substituting logic 1's for the *A* and *B* symbols, we obtain $1 + 1 = 0$, without a carry output from the INHIBITOR circuit. As shown, a separate line must be used to produce the carry digit. Hence, this circuit is known as a *half adder* and is useful for designing full binary adders for digital systems, as well as for code conversions, as more fully explained in Chap. 5. The truth table for the EXCLUSIVE OR is:

A	$+ B$	C
0	0	0
1	0	1
0	1	1
1	1	0

Another EXCLUSIVE OR arrangement is shown in the lower drawing of Fig. 3-11. While this has the disadvantage of an additional logic circuit it serves to illustrate the variations possible for obtaining the same end result. For an *A* input alone no inhibiting signal is present at the first INHIBITOR and an output is applied to the OR circuit, which in turn provides an output signal. For a *B* input alone, the second INHIBITOR applies a signal to the second input of the OR circuit and again an output is obtained from the EXCLUSIVE OR circuit. In neither case will an output be obtained from the AND switch because both *A* and *B* must be present here.

When both the *A* and *B* inputs are applied simultaneously both INHIBITORS are activated (logic *A but not B*) and no output is obtained from either one. Thus, no output is procured from the OR switch. With both *A* and *B* inputs applied, however, the AND circuit is functional and produces the *AB* output. Thus, the logical expression for this EXCLUSIVE OR circuit is $A\overline{B} + \overline{A}B$, which states that the true condition is *A but not B, or B but not A*. While this appears to be different than the expression $(A + B)(\overline{AB})$ obtained for the upper EXCLUSIVE OR circuit, the logic is the same (*A or B, but not A and B*).

The logic map and Venn diagram for the EXCLUSIVE OR circuit are shown in Fig. 3-12. For the map representation we again use the lower horizontal section to represent the logic 1 for *A* (true state) and the vertical right section for logic 1 for *B*. Since these intersect

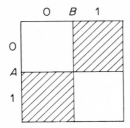
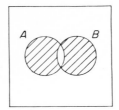

Fig. 3-12. Map and diagram for $A\bar{B} + \bar{A}B$.

at the lower right square and represent a false statement, the area is unshaded.

Similarly, for the Venn diagram, the intersection of the circles is left blank to express the logic that if both A and B are present the statement is false, but A alone is true, or B alone is true.

Questions and Problems

1. Briefly explain what is meant by two-level logic circuitry.

2. If A, B, C, are applied to a two-level circuit using negative logic and the expression is $(AB) + C$, would the expression change for a similar circuit using positive logic? Explain.

3. Draw a Venn diagram for the logic expression $(A + B) \cdot C$.

4. Draw a Venn diagram for the logic expression $(AB) + C$.

5. Explain what is meant by a matrix and also a logic tree.

6. What are the advantages of using transistors instead of diodes in two-level logic circuits?

7. What is the equivalent expression for $\overline{(A + B)(C + D)}$?

8. What is the equivalent expression for $\overline{(AB + C) + \overline{(DE)}}$?

9. What is the equivalent expression for $\overline{(A + \bar{B})(\bar{C} + \bar{D})}$?

10. Reduce the expression $\overline{(\bar{A} + \bar{B})} + \overline{(C \cdot D)}$ to its simplest form.

11. Simplify the expression $(A + B)(\overline{\bar{A} \cdot \bar{B}})$.

54

12. Simplify the expression $(A + B) + (A + C) + (A + B + C)$.

13. Write the resultant logic expressions for the symbolic diagrams in Fig. 3-13.

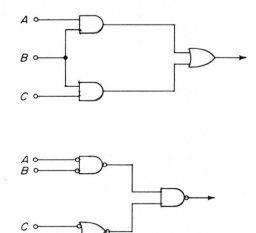

Fig. 3-13. Diagrams for Problem 13.

14. Write the resultant logic expressions for the symbolic diagrams in Fig. 3-14.

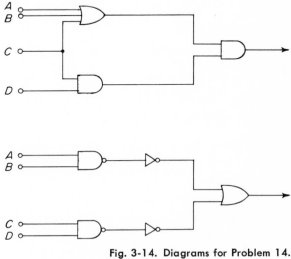

Fig. 3-14. Diagrams for Problem 14.

55

15. Write the resultant logic expression for the symbolic diagram in Fig. 3-15.

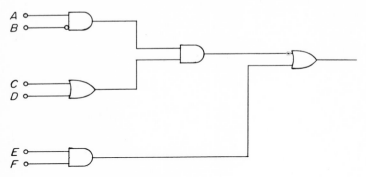

Fig. 3-15. Diagram for Problem 15.

16. Draw a symbolic diagram of a logic switching network to produce the expression $(A\bar{B}) + (CD)$.

17. Redraw the lower diagram of Fig. 3-10 without the NOT circuits and substitute any other symbols required to obtain the same output logic expression.

4

BOOLEAN ALGEBRA PRINCIPLES

INTRODUCTION

George Boole, as mentioned in the Introduction to Chap. 1, laid a foundation for a method of expediting the formation of efficient switching systems by formulating a logic-associated algebra now known as *Boolean algebra*. In 1854 he published *An Investigation of the Laws of Thought on Which Are Founded the Mathematical Theories of Logic and Probabilities*. In that text Boole proposed mathematical rules for producing logical conclusions by combining certain algebraic statements or propositions. Included were propositions for transforming statements into logic symbols, and using specific rules for manipulating the symbols to obtain logically valid solutions.

Later, additional contributions to symbolic logic concepts were made by two English mathematicians and writers, Alfred North Whitehead and Bertrand Russell. In 1910 they published *Principia Mathematica* which expounded the interrelationships between basic mathematics and formal logic, with deductions performed along strict symbolic principles. As such it also covered new areas in philosophy, particularly in *semantics* (the science of word meanings and word forms in language). More than 25 years were to elapse, however, before the value, significance, and practical applications of symbolic logic were to be recognized in relation to electric and electronic switching systems.

It was Claude E. Shannon who made a truly outstanding contribution to symbolic logic when he wrote his thesis for a Master of Science degree at the Massachusetts Institute of Technology. An abstract of this thesis entitled *A Symbolic Analysis of Relay and Switching Circuits* was published in the *Transactions of the American Institute of Electrical Engineers* in 1938. In this thesis Shannon investigated the methodology used in finding the most simple and effective switch combinations to obtain a desired result. The basic technique consisted of representing such complex multiswitch circuits by prescribed mathematical expressions, and providing the necessary arithmetical means for manipulating them based on the symbolic logic indicated by Boolean algebra. Thus, he unified the seemingly abstract symbolic logic with practical application in digital computers, telephone switching systems, and other electric and electronic applications or systems.

This chapter covers the basic principles of Boolean algebra, which represent the noteworthy contributions of many others, including Augustus De Morgan (1806–1871) an English writer, mathematician, and logican, who formulated one of the important theorems utilized in modern switching algebra.

THEOREMS, PRINCIPLES, AND POSTULATES

As discussed in preceding chapters, we have two logical connectives in switching circuitry. The plus sign ($+$) is used for the OR function and, as shown in Fig. 1-3, represents a parallel-circuit equivalent. The multiplication sign (raised dot or adjacent symbols) denotes the AND function and can be compared to the series circuit switches. In Boolean algebra we do not use the subtraction sign ($-$) or the division signs ($\div$ or $/$) since these functions are not valid or usable in logic expressions.

Letter symbols such as A, B, C, etc. are used in Boolean algebra to represent *variables*, each of which is called a *literal* symbol. Because such variables can only have a value of 1 or 0, proof of theorems becomes simple because only two values need be substituted for the variable. This proof procedure would be virtually impossible in ordinary algebra where variables may have thousands of different values.

In Boolean algebra, as in other mathematical forms, we have certain theorems, principles, and postulates from which we can deduce other laws or corollaries (assumed consequences inferred from valid and proved propositions). Using these various laws, theorems, and postulates, permit us to reduce complex switching circuitry by algebraic manipulations of Boolean expressions. The various laws and postulates of Boolean algebra necessary to this purpose follow.

Laws of Identity

For the OR logic, the statement $A + 0 = A$ indicates that A is true when applied to one of the inputs of a two-input OR switch, even though a 0 is present at the other input. With A having an assigned "1" value, we have the valid statement or the logic OR function: $1 + 0 = 1$. The input symbols could have been reversed, with the same end result: $0 + A = A$ as well as $0 + 1 = 1$.

For the AND logic, however, we must have coincidence at the inputs to obtain an output, hence $A \cdot 1 = A$. Representing the symbol A with a 1, we have: $1 \cdot 1 = 1$ which is a valid statement for the logic AND function. Similarly, $1 \cdot A = A$ as well as $1 \cdot 1 = 1$. Hence, 0 with respect to the OR function, and 1 with respect to the AND function are called *identity elements* for the logical connectives (operator $+$ as well as $\cdot$), because their usage leaves the A function unchanged. Thus, we can state two laws of identity:

The *addition* of 0 to a literal leaves the literal unchanged.
$(A + 0 = A)$ $\hfill$ (4-1)

The *multiplication* of a literal by 1 leaves the literal unchanged. $(A \cdot 1 = A)$ $\hfill$ (4-2)

Since these logic concepts have been covered in previous chapters, they present no difficulties in comprehension. These principles, however, as well as some others that follow, must be reemphasized and labeled in organized form to provide a solid foundation for proficiency in Boolean algebra manipulations.

Note the use of the terms *addition* and *multiplication* in laws (4-1) and (4-2), for the OR and AND functions. While it is true that the $+$ and $\cdot$ signs are logical connectives used with variables having only logical significance, it will be found that *some* of the postulates of Boolean algebra relate the $+$ and $\cdot$ operators to their true meaning in numerical algebra. Thus, it is convenient to use the terms *addition* and *multiplication* to indicate the OR ($+$) function as well as the

AND ($\cdot$) function. Consequently, the term *product* will also be used to identify the resultant expression of the AND operation, and *sum* to identify the resultant expression obtained by use of the OR operation.

Thus, A is the sum of the logical expression $A + 0$ and A is the product of the logical expression $A \cdot 1$. If we have an expression involving *A or B and A or C*, then $(A + B)(A + C)$ is called the *product of sums*. Similarly, if the expression is *A and B or C and D*, we state that $AB + CD$ is the *sum of products*. The use of these terms simplifies explanations of laws, theorems, and postulates, and no confusion should arise if we keep in mind that any variable can have a logical significance of only 0 or 1, and that the $+$ sign is a logical connective representative of the OR function and the $\cdot$ sign is a logical connective representative of the AND function. The coincidental relationship of these operators to their meaning in numerical algebra will become more evident in subsequent discussions.

Laws of Complementation

The basic concepts of the NOT function (logic negation) were introduced in Chap. 1, and it was pointed out that the *not* logic is also referred to as a *complementing* function. If $A = 1$, then $\bar{A} = 0$, and a statement such as $A \cdot \bar{A} = 0$ indicates that A and NOT A equal zero, or $1 \cdot 0 = 0$. Negation of an expression involves the literals as well as the operators (logical connectives). Hence, to complement an expression, we must:

Change all 1's to 0's; change all 0's to 1's; change each
logical connective AND to OR; change each logical connec-   (4-3)
tive OR to AND; and change each A to $\bar{A}$; $\bar{A}$ to A; B to $\bar{B}$; etc.

Thus, the complement of the expression $A + (0 \cdot \bar{B})$ becomes $\bar{A}(1 + B)$, and the complement of the expression $(\bar{A} + B) \cdot 1$ becomes $(A\bar{B}) + 0$.

Law of Involution

In Chap. 1 it was also pointed out that a double negation restores a logic expression to its original form. Thus, $\bar{\bar{A}} = A$; $\bar{\bar{1}} = 1$; etc. Similarly, if we complement the expression $A + B$ we get $\overline{A} \cdot \overline{B}$. If we now complement *this* expression we obtain $\overline{\overline{A}} + \overline{\overline{B}}$, which restores the logic expression to the original $A + B$. Our law of involution, therefore, is:

If the complement is performed twice, the original function is regained. ($\overline{\overline{A}} = A$.) (4-4)

Law of Dualization

In Boolean algebra, a *dual expression* is obtained by negating 0's, 1's, and the logical connectives, but not any literals. Thus, $A + (0 \cdot \overline{B})$ becomes $A(1 + \overline{B})$ and forms a dual pair with the original expression. Note that this is not the complement expression $\overline{A}(1 + B)$. This duality of Boolean expressions is an important factor in logic algebra, and the law is expressed as:

If we change identity elements (1, 0) and the operators (logical connectives), but do not form complements of literals, a duality is achieved and the statement validity does not change. (4-5)

The significance of the law of dualization will become more apparent as we examine De Morgan's theorem, which follows, where the dual pair is related to the law of equivalence. The dualization law is also of value in proving theorems, since the proof of one expression of a dual identity also applies to the other, and hence additional proof is unnecessary.

DE MORGAN'S THEOREM

An important theorem was developed by Augustus De Morgan which has proved of considerable value in Boolean algebra. It is useful for ascertaining the complement of any Boolean expression and relates to the *law of equivalence*. The concept of this theorem can be more readily achieved by first investigating the basic difference in logic when a negation precedes a logic expression, and when a negation follows an expression.

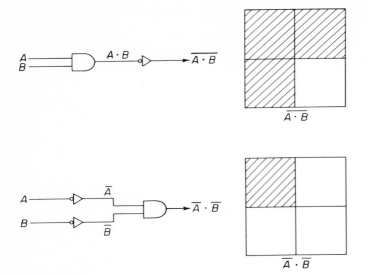

Fig. 4-1. Effects of NOT function on AND logic.

The effect of the NOT function on AND logic is shown in Fig. 4-1. Note the difference in the Boolean statement that results in an output use of the NOT circuit compared with that of having a NOT circuit preceding a logic gate. As shown for the upper AND circuitry, when the NOT circuit is in series with the output, the $A \cdot B$ value is inverted and the output expression becomes $\overline{A \cdot B}$ as shown, with the logic map indicating the function. When the NOT circuit is in series with the input of the AND switch as shown in the lower portion of Fig. 4-1, the A and B quantities are inverted before application to the AND circuitry and hence $\overline{A}$ and $\overline{B}$ values are applied to the inputs. The resultant output is then $\overline{A} \cdot \overline{B}$. This logic function is also illustrated with the logic map at the right of the symbolic representation. (See also Fig. 2-5.)

Note that the expression $\overline{A \cdot B}$ also negates the logical connective, and since the complement of the AND function is the OR function, the expression is equivalent to $\overline{A} + \overline{B}$. For $\overline{A} \cdot \overline{B}$, however, the connective is not negated.

Similar differences occur between the condition where a NOT circuit follows an OR switch and where the NOT circuit precedes the OR switch, as shown in Fig. 4-2. When the NOT circuit negates the output as shown in the upper symbolic representation in Fig. 4-2,

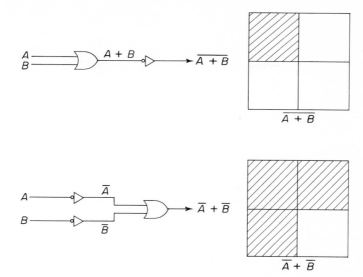

Fig. 4-2. Effects of NOT function on OR logic.

and $A + B$ is complemented; the result is $\overline{A + B}$. For the lower representation, the NOT circuits are in series with the individual A *or* B inputs, and hence negated variables are applied to the OR gate, resulting in an output of $\overline{A} + \overline{B}$ as shown. The appropriate logic maps indicate the representative functions. Note that the logic map for $\overline{A + B}$ in Fig. 4-2 is identical to the logic map for the expression $\overline{A} \cdot \overline{B}$ in Fig. 4-1. Also note that the logic map for $\overline{A} + \overline{B}$ of Fig. 4-2 is exactly the same as the map for the expression $\overline{A \cdot B}$ of Fig. 4-1. This illustrates De Morgan's theorem, or the law of equivalence:

The complement of a product of literals is equivalent to the sum of the separate literal complements; and the complement of a sum of literals is equivalent to the product of the separate literal complements.   (4-6)

A comparison of the equivalency of the four logic functions is shown in Fig. 4-3, indicating that $\overline{AB} = \overline{A} + \overline{B}$ and that $\overline{A + B} = \overline{A} \cdot \overline{B}$. Note that equivalence is involved here, and not the complement of an expression. Hence, $\overline{A} + \overline{B} = \overline{AB}$ relates to the *law of equivalence*, while the complement of $\overline{A} + \overline{B}$ produces $A \cdot B$, and a dual pair is formed [law of dualization (4-5)] by $\overline{A} \cdot \overline{B}$.

63

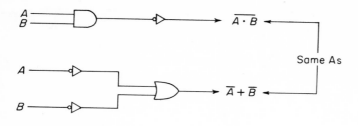

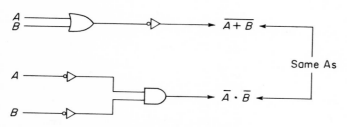

Fig. 4-3. Logic illustrations of De Morgan's theorem.

For the equivalence shown in Fig. 4-3, it should be noted that the negation of the AND $(\overline{A \cdot B})$ function (NOT AND) is equal to the alternate denial $(\overline{A} + \overline{B})$, which expresses, in essence, that *not A or not B* is true. Similarly, the negation of the OR $(\overline{A + B})$ function (NOT OR), is equal to the joint denial $(\overline{A} \cdot \overline{B})$, which states that *not A and not B* is true. The following truth tables establish the validity of De Morgan's theorem:

$A \cdot B = (A \cdot B)$			$(\overline{A \cdot B})$
0 0	0	negation =	1
0 1	0		1
1 0	0		1
1 1	1		0

$\overline{A} + \overline{B} = (\overline{A} + \overline{B})$		
1	1	1
1	0	1
0	1	1
0	0	0

same results, hence:

$$\overline{A \cdot B} = \overline{A} + \overline{B}$$

These truth tables apply to the upper drawings of Fig. 4-3 and show the equivalence logic. The upper truth table shows the results obtained when the output from an AND switch is fed to a NOT circuit. The entire $A \cdot B$ output expression is negated and results in $\overline{A \cdot B}$ as shown, which includes the negated logical connective. In the lower truth table, a NOT circuit precedes each OR switch input, resulting in an output of $\overline{A} + \overline{B}$ as shown. (Note that the input values shown in the lower table are the complements of the inputs in the upper table—representing the NOT function input.) Because the individual output values for $\overline{A \cdot B}$ are the same as those for $\overline{A} + \overline{B}$, De Morgan's theorem is validated.

The following truth tables apply to the lower illustrations of Fig. 4-3. The first table shows the results obtained when the output from an OR switch is fed to a NOT circuit. The $A + B$ output is negated and results in $\overline{A + B}$ as shown, which also includes the negated logical connective. For the second truth table, a NOT circuit precedes each AND switch input, hence the individual input values are complements of those in the first table. The resultant output is $\overline{A} \cdot \overline{B}$ as shown. Again, the individual *output* values for $\overline{A + B}$ are the same as those for $\overline{A} \cdot \overline{B}$, again validating De Morgan's theorem.

$$A + B = (A + B) \qquad\qquad \overline{(A + B)}$$

0	0	0	negation =	1
0	1	1		0
1	0	1		0
1	1	1		0

$$\overline{A} \cdot \overline{B} = (\overline{A} \cdot \overline{B})$$

		same results,
1 1	1	hence
1 0	0	$\overline{A + B} = \overline{A} \cdot \overline{B}$
0 1	0	
0 0	0	

De Morgan's theorem also applies to switches with multiple inputs. Thus, a three-input AND switch, with NOT circuits in series with each input, has an output of $\overline{A} \cdot \overline{B} \cdot \overline{C}$ which is equivalent to a three-input, OR switch followed by a NOT circuit to produce $\overline{A + B + C}$. Similarly, $\overline{ABC}$ is equivalent to $\overline{A} + \overline{B} + \overline{C}$, etc. Again, the process

65

must not be confused with the complement function, because the complement of $\overline{A} + \overline{B} + \overline{C}$ is ABC. Similarly, the expression $\overline{A \cdot \overline{B} \cdot \overline{C} \cdot D}$ is equivalent to $\overline{A} + B + C + \overline{D}$, but the complement of the latter expression is $A \cdot \overline{B} \cdot \overline{C} \cdot D$.

De Morgan's theorem in Boolean switching algebra is valuable because it permits all AND logic to be changed into OR logic (or vice versa). Since the complement of any logic expression usually can be formed easily, De Morgan's theorem becomes useful for changing the expression form to help simplify switching circuitry.

Some Boolean algebra axioms form equation statements unlike those found in numerical algebra. In ordinary algebra, for instance, the expression $A \cdot A$ results in A^2, but in Boolean algebra we have no exponents, and our coefficients are confined to only 1 and 0. Thus, the expression $A \cdot A$ (as well as $A + A$) gives us an A for an answer:

$$A \cdot A = A \qquad (4\text{-}7)$$

$$A + A = A \qquad (4\text{-}8)$$

Thus, for the AND function the power is not raised, nor is the A value increased for the OR logic. Actually the power remains the same and hence this axiom is often known as the *law of idempotency*. The validity of the axiom is illustrated in Fig. 4-4. Here, the $A \cdot A$ expression is shown as two switches in series. Because both are labeled A each has the same circuit- and switching-characteristic functions as the other and can be replaced by a single switch as shown. Thus, this logical expression is actually redundant, because if A *and* A are indicated, one can be dispensed with. Similarly, the $A + A$ expression can be represented as a parallel circuit, as shown in the second drawing of Fig. 4-4. Again, if A is needed, then the other OR function involving the second A is unnecessary, and the single switch A is adequate.

The law of idempotency leads us to the simplification of other expressions, and the A symbols in Eqs. (4-7) and (4-8) are only representative. If other symbols are used in duplication, the law of idempotency still applies. This is illustrated in the lower drawing of

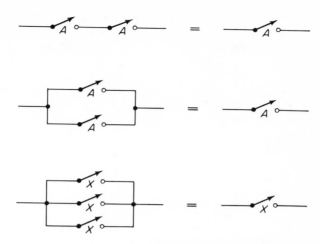

Fig. 4-4. Idempotency law logic.

Fig. 4-4, where the expression $X + X + X$ is equal to a single switch X as shown.

Other expressions are illustrated in Fig. 4-5. The upper drawing represents the expression $A(A + B) = A$. From inspection we can

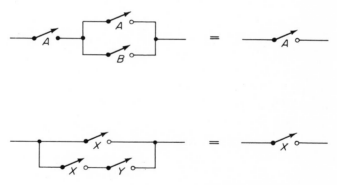

Fig. 4-5. Switching analogies.

see that this three-switch circuit is equivalent to the simple expression A as shown. Again, if A is required, the other OR expression containing A is redundant. Similarly, $X + (X \cdot Y)$ is equal to X as shown in the second drawing of Fig. 4-5, because the AND switch involving X is again unnecessary.

67

Thus, any expression such as $A \cdot (A + B + C + D)$ can be simplified by replacing it with A in Boolean algebra. Similarly,

$$X + (X \cdot Y \cdot Z) = X.$$

If, originally, we had encountered these in combined form as a single expression $A \cdot (A + B + C + D) \cdot [X + (X \cdot Y \cdot Z)]$, the simplified version would be $A \cdot X$. As an additional example consider the expression $A \cdot (A + B) [Z + (Y \cdot Z) \cdot \overline{W}]$. The initial expression $A(A + B) = A$, and the second member $(Z + YZ) = Z$. Combining these, and adding them to the remaining symbol $\overline{W}$, we have $A \cdot Z \cdot \overline{W}$ as our simplified resultant.

Negated literals may also be cancelled out when the law of idempotency is involved. Thus, $(A\overline{B}) + (AB) = A$. In the first part of the expression we are stating, in essence, that A *but not* B is true, or that A alone is true. Hence, $A + (AB) = A$. Had the expression been stated as $(AB) + (A\overline{B})$ the end result would still have been an A.

Similarly, when we apply the law of idempotency to the expression $(A + B)(A + \overline{B})$ we again obtain A. In this expression we are adding A to the product of $B\overline{B}$: $A + B\overline{B}$. Since $B \cdot \overline{B} = 1 \cdot 0$, the product is 0. When this is added to A we have $A + 0 = A$.

The logic of the *but not* must be observed in an expression such as $A + \overline{A}B$ or $A + B\overline{A}$. Here we state that A is true *or* B but not A. Hence, A is true *or* B *alone* is true and the result is $A + \overline{A}B = A + B$. Similarly, $A(\overline{A} + B) = AB$, because it states that A *and* B but not A is true, or that A *and* B is true.

Notice the absence of parentheses in most expressions in the preceding paragraph. When this is done we follow the practice of ordinary algebra where $A + BC$ would mean $A + (BC)$ and not $(A + B)C$.

Various expressions involved with the law of idempotency are listed below for reference:

$$A \cdot 1 = A$$
$$A \cdot A = A$$
$$A \cdot \overline{\overline{A}} = A$$
$$A \cdot \overline{A} = 0$$
$$A \cdot 0 = 0$$
$$A + A = A$$
$$A + \overline{A} = 1$$
$$A + \overline{\overline{A}} = A$$

$$A + 0 = A$$
$$A + 1 = 1$$
$$A(A + B) = A$$
$$A + (A \cdot B) = A$$
$$(A\bar{B}) + (AB) = A$$
$$(A + B)(A + \bar{B}) = A$$
$$A(\bar{A} + B) = AB$$
$$A + (\bar{A}B) = A + B$$

COMMUTATIVE LAW

Boolean algebra uses several principles that parallel laws found in ordinary arithmetic and algebra. Besides the *laws of identity* covered earlier, we also utilize the *commutative law,* the *associative law,* and the *distributive* law. The *commutative law* emphasizes the fact that the order in which addition or multiplication is sequenced has no effect on the sum or product. Hence, $A + B = B + A$ for the OR function, and $AB = BA$ for the AND function. We can express the law formally as

The order in which we add or multiply a pair of variables n does not alter the end results.   (4-9)

Note that we use the terms *add* and *multiply* for convenience in designating the operations of the logical connectives for the OR and AND functions, as explained earlier in this chapter in the section on *Laws of Identity.*

The validity of the logic applying to the commutative law is illustrated in Fig. 4-6. At the top is shown the OR circuit symbol with an $A + B$ input. By the commutative law for addition we get an output regardless of the interchange of inputs and $A + B = B + A$ as shown. The comparable physical switch arrangements are shown next, with parallel circuits representing the OR function. Again, either A or B closes the circuit.

The AND circuit symbols represent the commutative law for multiplication and the equivalent switch representations are shown at the bottom. As with the OR circuit logic, the A and B representations can be interchanged with the same result, $AB = BA$.

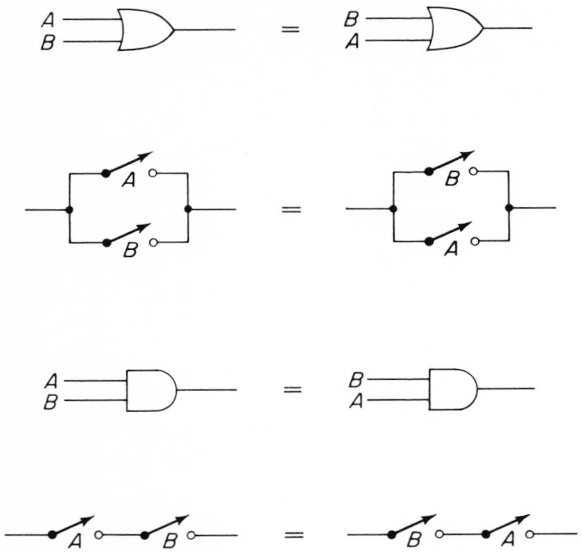

Fig. 4-6. Commutative law logic.

ASSOCIATIVE LAW

The associative law in Boolean algebra indicates that the variables of a Boolean expression may be grouped as desired so long as they are connected by the same logical operator. Thus,

$$(A + B) + C = A + (B + C) = A + B + C,$$

and this represents the associative law for addition. For multiplication we have $(AB)C = A(BC) = ABC$. The law can be expressed formally as:

The order in which variables n are added or multiplied does not alter the end result.　　　　　　　　　　　　　　　　　　(4-10)

The logical validity can again be demonstrated by using symbols of OR and AND or by switch schematics as shown in Fig. 4-7. The logic-symbol representation of the expression $(A + B) + C$ is shown at the upper left. Because the $A + B$ inputs are applied to an individual OR circuit they represent the $(A + B)$ portion of the expression. This arrangement is the same as the upper center one where the $(B + C)$

70

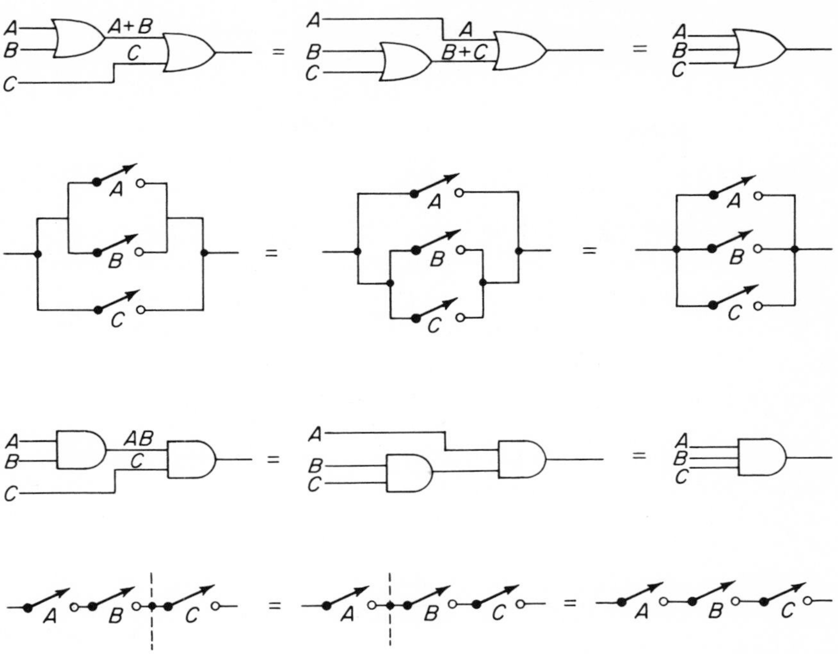

Fig. 4-7. Associative law logic.

inputs are applied to a single OR circuit and hence represents the parenthetical portion of $A + (B + C)$. Because the logic is A or B or C, the simplified equivalent is shown at the upper right as a three-input OR switch.

Next are shown the equivalent physical-switch arrangements where the $(A + B)$ portion of the expression is shown as a parallel-switch entity. Again, the end result is a simple three-switch parallel arrangement as shown at the right.

The logic for the associative law of multiplication is shown in the lower drawings of Fig. 4-7. The two inputs to the first AND switch form the (AB) portion of the expression $(AB)C$ which is the same as using the (BC) inputs to the first AND circuit as shown at the center. Again, these are the same as using a single logic circuit. In this instance a three-input AND switch suffices, as shown at the lower right. The equivalent switch arrangements are shown at the bottom with the AND function represented by series switches. The parenthetical expressions are represented by the grouping of two switches to one

71

side of the dashed vertical line. These are identical to the simple series switch arrangement shown at the lower right.

DISTRIBUTIVE LAW

In ordinary algebra the distributive law is exemplified by the equality of the expressions $A(B + C) = (AB) + (AC)$. This can be proved easily by substituting numerals such as $A = 2$, $B = 3$, and $C = 6$. Thus, $2 \cdot (3 + 6) = 2 \cdot 9 = 18$. Substituting these values in $(AB) + (AC)$ yields: $(2 \cdot 3) + (2 \cdot 6) = 6 + 12 = 18$. This operation is also valid in Boolean algebra, and the law can be expressed as:

Multiplying the sum of two variables *n* by another variable
n is the same as adding the product of the first and second   (4-11)
to the product of the first and third.

In Boolean algebra, however, another distributive law is used which states: $A + (BC) = (A + B)(A + C)$. If we substitute our values $A = 2$, $B = 3$, and $C = 6$ for these expressions, we would find that the equality indication for the two expressions does not hold in numerical algebra: $2 + (3 \cdot 6) = 2 + 18 = 20$. For the $(A + B)(A + C)$ form we would get: $(2 + 3)(2 + 6) - 5 \cdot 8 = 40$, which is not the same as our answer for the expression $A + (BC)$. For *Boolean algebra*, however, *the logic is valid*, since the first expression states that *A or B and C* is a true statement. Hence, *A* alone is true, as is *A and C*, *or B and C*, or the total expression *(A or B) and (A or C)*. Hence, we can state the law in formal terms as:

Adding a variable *n* to the product of two other variable *n*
values is the same as multiplying the sum of the first and   (4-12)
second by the sum of the first and third.

The symbol representations of Fig. 4-8 will help clarify the equalities representative of the distributive law. For the drawing at the upper left, the two inputs to the OR switch form the parenthetical expression $(B + C)$ which, in conjunction with the A input, provides for the output $A(B + C)$. The latter, in turn, is equal to the representation at the upper right showing the $(AB) + (AC)$ diagram.

For this upper-right drawing, note that A alone is not true, because

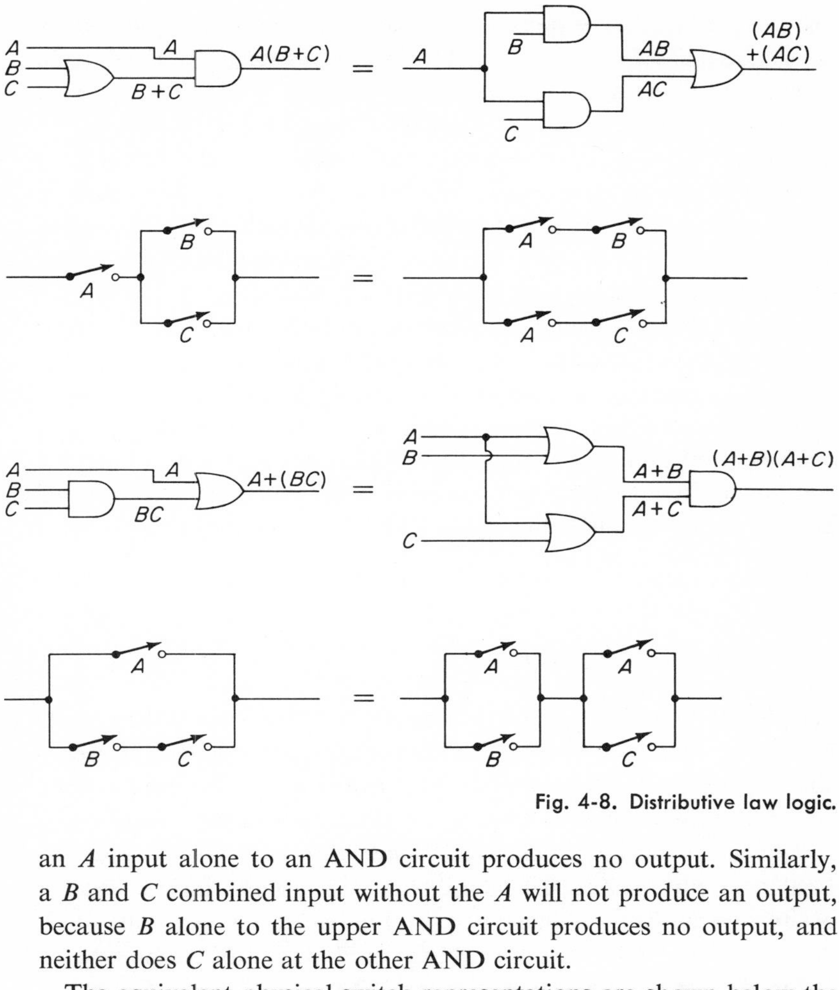

Fig. 4-8. Distributive law logic.

an A input alone to an AND circuit produces no output. Similarly, a B and C combined input without the A will not produce an output, because B alone to the upper AND circuit produces no output, and neither does C alone at the other AND circuit.

The equivalent physical-switch representations are shown below the symbolic drawings. For the left arrangement, it is obvious that A must be closed *and* either B *or* C. Thus, the logic for the switches at the right is identical, since A *and* B is true, *or* A *and* C is true.

The logic-gate symbolic representation of $A + (BC)$ is shown at the lower left of Fig. 4-8. The two inputs to the AND switch form the parenthetical expression (BC) applied to the OR switch. Note that when we use both A and B we obtain an output since A alone would provide one. The B would not open the AND switch (nor would C alone). Since A provides an output, however, the AND gate need not open. Since either *A or B and C* produces an output, the circuit logic is the same as that shown at the right for the expression $(A + B)(A + C)$, where *A or B and A or C* provides a true output.

The equivalent circuits for mechanical switches are shown at the bottom of Fig. 4-8. For the circuit at the bottom left, either *A or B and C* will permit the passage of signals and hence is represented by the expression $A + (BC)$. The same function is realized by the switching arrangement at the bottom right, where both switches designated as *A* must be closed, *or* both lower switches must be closed (*B and C*). A true statement still prevails, however, because *A and C*, B *and* C, as well as *A* or *B and A or C* executes the required logic.

Thus, by use of the distributive law, logic expressions such as $(AB) + (AC)$ and $(A + B)(A + C)$ can be reduced to their logical equivalent of $A(B + C)$ and $A + (BC)$ and thereby eliminate the additional circuitry otherwise required. One logic gate is eliminated in each circuit shown at the right in Fig. 4-8, by using the equivalent circuitry shown at the left.

NEGATIVE LOGIC AND NEGATION

Chapter 2 (Positive and Negative Logic) pointed out that *negative logic* must not be confused with the *logic of negation*. This point is reemphasized here since the logic-symbol representations for Boolean algebra laws and theorems have been shown in positive-logic form. Equivalent circuits using negative-logic symbology were not shown, since the *logic* is not altered if a particular circuit uses negative logic instead of positive logic. This is illustrated in Fig. 4-9. The logic circuit at the top for $A + (BC)$ could have been used instead of the one shown in Fig. 4-8 for this Boolean expression. The only difference is that the symbolic representation in Fig. 4-9 indicates that negative signals represent the true (or 1) state. Since this would be of academic interest only (in terms of hardware design), it serves no purpose to show negative logic in preference to positive logic when showing symbolic representations of Boolean-algebra switching systems.

Negative logic is not logical negation, because the variables *A* or *B* could be represented by positive or negative signals, as desired. If positive signals are used, *A* and *B* represent "one" quantities, and $\overline{A}$ and $\overline{B}$ represent "zero" quantities. Similarly, if negative signals are

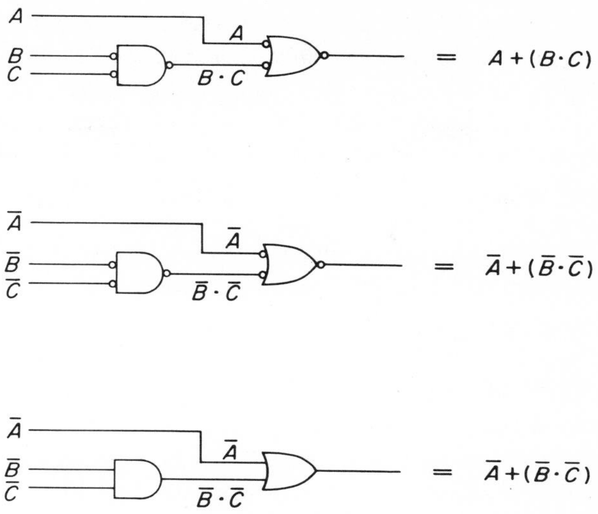

$$= A + (B \cdot C)$$

$$= \overline{A} + (\overline{B} \cdot \overline{C})$$

$$= \overline{A} + (\overline{B} \cdot \overline{C})$$

Fig. 4-9. Negative logic and negation

used for logic 1's, the A and B symbols represent "one" quantities again, and $\overline{A}$ and $\overline{B}$ indicate "zero" quantities.

The second drawing of Fig. 4-9 shows the NOT variables represented by $\overline{A}$, $\overline{B}$, $\overline{C}$. The fact that these are shown for negative-logic circuits has nothing to do with the logical expressions themselves. This is exemplified by the lower drawing which shows positive logic, but with negated variables for the inputs. Note that the logic that applies is no different from that for the negative-logic circuit also using negated variables for the inputs.

The laws previously discussed apply to the complemented (negated) expressions also. The complement of the expression $A + (BC)$, for instance, is $\overline{A}(\overline{B} + \overline{C})$ and by the distributive law (4-11), this is the same as $(\overline{A} \cdot \overline{B}) + (\overline{A} \cdot \overline{C})$. If we complement the expression $A(B + C)$ the negation yields $\overline{A} + (\overline{B} \cdot \overline{C})$ which, according to distributive law (4-12), is the same as $(\overline{A} + \overline{B})(\overline{A} + \overline{C})$.

Questions and Problems

1. What two mathematical signs are not used in Boolean algebra?

2. What mathematical law applies to the Boolean algebra expressions $0 + A = A$; $A \cdot 1 = A$?

75

3. Using W, X, Y, and Z for variables, show expressions representing *product of sums* and *sum of products*.

4. Show the complements of the expressions:

$$\bar{A} + (1 \cdot B); \; (A + B)(A + C); \; A + \bar{A}B; \text{ and } (0 + \bar{B}) \cdot A$$

5. Briefly explain what is meant by the *law of involution*.

6. What expression forms a dual pair with $\bar{A} + (0 \cdot B)$ and what is the complement of each expression of the dual pair?

7. What is the difference between the expression $\overline{A + B}$ and $\bar{A} + \bar{B}$?

8. Briefly explain De Morgan's theorem.

9. Indicate which two expressions of the following group relate to the law of equivalence, which two relate to the law of complementation, and which to the law of dualization:

$$\bar{A} \cdot \bar{B} \quad A + B$$
$$\bar{A} \cdot \bar{B} \quad \bar{A} + \bar{B}$$
$$\overline{A \cdot B} \quad \overline{A + B}$$

10. Briefly explain what is meant by the *law of idempotency*.

11. Simplify the expressions: $A(A + \bar{B})$; $\; A + (A \cdot B \cdot C)$; and $A(A + B + C + D + E)$.

12. Simplify the expressions: $(X\bar{Y}) + (XY)$; $(A + \bar{B})(A + B)$; and $(A + B)(A + \bar{B})(A + C)$.

13. Briefly explain what is meant by the *commutative law*.

14. Briefly explain what is meant by the *associative law*.

15. Define, in your own words, the *distributive law* for addition.

16. Which distributive law does not hold for numerical algebra?

17. Reduce $(\bar{A} + \bar{B})(\bar{A} + \bar{C})$ to its logical equivalent expression.

18. Simplify the expression $(XY) + (XZ)$.

19. Complement the expression $X(Y + Z)$ and apply the distributive law to the complemented expression to show the logical equivalent.

20. Complement the expression; $(\bar{A} \cdot \bar{B}) + (\bar{A} \cdot \bar{C})$; then use the distributive law to simplify the complemented expression.

5

NUMBERS, ADDERS, AND CODES

Introduction

Logic switching circuits are used not only in telephone systems and industrial control circuitry, but are also extensively employed in modern digital computers. Not only are the "on" and "off" states utilized exclusively, but their equivalent "1" and "0" states are integrated with the binary numbering system to perform various arithmetical tasks.

The study of Boolean algebra is facilitated by an understanding of the binary numbering system, its codes, and the manner in which they are associated with switching systems. Also of help is a knowledge of certain definitions applied to the numbering system and codes, particularly in terms of their relationship to Boolean algebra usage, as covered in this chapter.

Binary Numbers

The aggregate of *digits* employed in a particular numbering system is referred to as the *radix*. Thus, radix 10 designates the base-ten system wherein we have 10 digits ranging from 0 to 9. When we employ the "on" and "off" switching states to form a numbering system containing only 0 and 1, it represents radix 2 and is also known as the *binary* system.

The binary numbering system is related to the powers of 2, and each power of 2 which a place represents is indicated by the following:

8	7	6	5	4	3	2	1	Place
2^7	2^6	2^5	2^4	2^3	2^2	2^1	1	Power
128	64	32	16	8	4	2	1	Value

Note that the value of each place doubles from right to left. Thus, the binary number 1 has a value of 1, but the binary number 10 has a value of 2, while the binary number 100 has a value of 4. Similarly, the binary number 101 has a value of 5, because the values of the first and third places, when added, equal 5. Also, if the binary number is 111, it represents 7 (4 + 2 + 1 values). Representative values from 0 to 20 are shown in the table below:

Binary Value	Decimal Value
00000	0
00001	1
00010	2
00011	3
00100	4
00101	5
00110	6
00111	7
01000	8
01001	9
01010	10
01011	11
01100	12
01101	13
01110	14
01111	15
10000	16
10001	17
10010	18
10011	19
10100	20

From inspection it is obvious that the table can be extended indefinitely by continuing the configuration peculiar to the binary system. Note that in the first-place column the numeral 1 *alternates* with 0 for the entire vertical column. In the second-place column, two 0's alternate with two 1's in progression down the entire length. In the third-place column, four 0's alternate with four 1's for the entire length. The same radix-2 factor prevails for all successive columns to the left, following the 1, 2, 4, 8, 16, 32, etc. characteristic pattern.

Binary numbers lend themselves to arithmetical manipulations just as with base-ten numbers. If, for instance, the binary number 11 (having a value of 3) is to be added to the binary number 100 (having a value of 4), the answer is binary 111 (7) as shown below:

$$
\begin{array}{rl}
11\ (3) & \text{addend} \\
+100\ (4) & \text{augend} \\
\hline
111\ (7) & \text{sum}
\end{array}
$$

Similarly:

$$
\begin{array}{lll}
101\ (5) & 1000\ (8) & 1000\ (8) \\
+010\ (2) & 0101\ (5) & 0001\ (1) \\
\hline
=111\ (7) & +0010\ (2) & +0100\ (4) \\
& \hline & \hline \\
& =1111\ (15) & =1101\ (13)
\end{array}
$$

Just as with the base-ten system, however, there are occasions where the addition process requires the "carrying" of a number. If, for instance, 01 is to be added to 101, it would be set down as follows:

$$
\begin{array}{r}
01 \\
+101 \\
\end{array}
$$

The addition of the two binary digits in the first-place column would produce the base-ten value of 2; but, since the numeral 1 is the highest number in the binary system, we encounter the rule that $1 + 1 = 0$ *with* 1 *to carry*. Thus, the addition of the two 1's in the first-place column produces a 0 and a binary 1 is then carried to the second place. Thus, the sum is 110 (6):

$$
\begin{array}{r}
01\ (1) \\
+101\ (5) \\
\hline
110
\end{array}
$$

Occasions arise, of course, when it is necessary to carry more than a single digit when several binary numbers are to be added:

$$
\begin{array}{r}
11\ (3) \\
11\ (3) \\
+\ \ 101\ (5) \\
\hline
1011\ (11)
\end{array}
$$

79

Here, the top first-place 1 is added to the next lower 1 which produces 0 and 1 to carry. The remaining 1 in the first-place column (plus the 0 resulting from the addition of the two 1's) equals 1. Hence the addition of the first-place column produces a 1. When the carry 1 is added to the top 1 in the second-place column, the result is again 0, with 1 to carry. The 0 added to the remaining 1 in the second-place column produces a 1, which is entered in the second place in the answer. Now the carry 1 is added to the third-place 1, producing a 0 with 1 to carry into the fourth place as shown. The validity of the answers can be checked by comparing the base-ten values of the binary numbers to be added to the equivalent binary number of the base-ten sum.

ADDERS

The manner in which logic-switching circuits are combined to perform addition with carry can be more readily understood by first investigating the characteristics of the half adder. The half adders are also known as EXCLUSIVE OR circuits and the applicable logic was discussed in Chap. 3, as distinguished from the INCLUSIVE OR circuit discussed in Chap. 2. Two typical half adders were illustrated in Fig. 3-11.

Half adders have no recirculating carry function and hence cannot perform complete addition. They are useful, however, for performing code conversions, as discussed more fully later in this chapter. A full adder of two serial-train binary numbers of any length is formed by combining two half adders with appropriate carry-recirculation loops, as shown later. The half adder, by itself, will produce a true sum only if no carry function is involved, as shown in Fig. 5-1A. Thus, with 101 applied to the A input of the OR circuit, and 010 to the B input, an output of 111 is produced from it. With 101 and 010 applied to the AND circuit, no coincidence prevails and hence no output is produced from it. Thus, 111 is applied to the A input of the inhibitor and 000 to the B input. Since no inhibiting function is initiated, the true sum output is obtained (111).

If a carry operation is involved a false sum will appear as shown in Fig. 5-1B. Here 101 is applied to the A inputs of the OR and AND circuits, and 001 to the B inputs. Hence, the output from the OR circuit is 101, while that from the AND circuit is 001 (since no coinci-

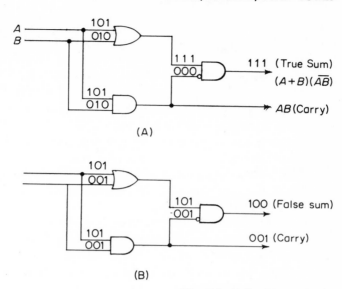

Fig. 5-1. Half-adder functions.

dence occurs for the third-place 1 in 101). The first-place 1 at the inhibiting input produces a 100 output from the half adder as shown instead of the true sum 110.

By combining two half adders and including a recirculation loop for the carry function, a full adder is formed, as shown in Fig. 5-2. Any carry digit that is recirculated undergoes a 1-place delay through a pulse delay line as shown. Diode D_1 prevents the recirculated carry digit from entering circuits other than the delay line. Diode D_2 prevents reverse circulation of any digits applied to the delay line from the first AND circuit (designated as A-1). The diodes as shown assume positive pulses represent logic 1 digits. If negative pulses indicate logic 1's the diodes would be reversed, with no change in the logic involved.

The full adder shown in Fig. 5-2 will accept two serial-train binary numbers of any length and produce a true-sum output. Two binary input numbers (01 and 11) are shown to illustrate the logic functions that apply. When the first-place digits appear at the first OR circuit (designated as 0-1) an output digit is produced and applied to the first inhibitor input (designated as I-1). For the two-digit input of the first AND circuit (A-1) coincidence occurs and a digit output is also produced. The output from the A-1 AND circuit is applied to the

81

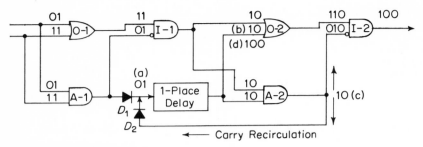

Fig. 5-2. Two-input full adder.

inhibiting input (I-1) as well as to the 1-place delay line, as shown.

For the first inhibitor (I-1) the first-place digits cancel and produce 0 output. Similarly, the 1-place delay produces 0 output. Hence no first-place 1's appear at the second OR circuit (0-2) or at the second AND circuit (A-2). Thus no digits enter the second inhibitor (I-2) and 0 output results from the adder.

The second-place digit at the input to the first OR circuit (0-1) applied a digit to the input of inhibitor I-1. For the single-digit input to A-1, however, no coincidence occurs and no output is produced. Since no inhibiting function occurs for I-1, an output digit is produced and applied to the inputs of both 0-2 and A-2 as shown. In second place, however, the delay line furnishes an output pulse because the 01 input identified as (a) in Fig. 5-2 appears as the (b) output 10. Now both the second OR circuit (0-2) and the second AND circuit (A-2) have coinciding digit inputs. Hence each produces an output digit in the second place and these are applied to the second *inhibitor* (I-2). Because the inhibiting function occurs, there is no second-place digit output. The second-place digit output from A-2 in the form of 10 and marked (c), is the carry digit. This is recirculated to the delay line and appears as a third-place digit 100 and identified as (d). This cannot enter the AND circuit (A-2) because there is no coinciding input here. The third-place digit at the second OR circuit produces an output, however, and since no inhibiting digit is present in third place at I-2, it appears at the output to produce the true sum of 100.

A full adder can be formed without recirculation of the carry digit, but with provisions for a carry input as well as a carry output, as shown in Fig. 5-3. Here, the addend and augend inputs are designated as A and B, and the carry input as C_i. The output sum is indicated

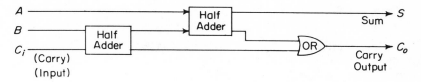

Fig. 5-3. Three-input serial adder.

as S and the carry output as C_o. The resulting truth table is shown below, with each logic expression row numbered for identification.

Row	Inputs			Outputs	
	A	B	C_i	S	C_o
1	0	0	0	0	0
2	0	0	1	1	0
3	0	1	0	1	0
4	0	1	1	0	1
5	1	1	0	0	1
6	1	0	0	1	0
7	1	0	1	0	1
8	1	1	1	1	1

The sum and carry outputs conform to the rule mentioned earlier in this chapter: $1 + 1 = 0$ with 1 to carry. Thus, for row 2 in the foregoing truth table, 1 added to $0 = 1$ as the sum, with no carry. Similarly, for row 3, the single digit input at B also indicates $1 + 0$, with no carry, and a 1 output. For row 4, however, the B input is added to the C_i (carry) input, producing a sum of 0, with 1 to carry at the C_o output. The same condition prevails in row 5, where inputs *at* A as well as at $B = 1 + 1 = 0$, with 1 to carry. Row 6 produces the same output as either row 2 or 3, since a single input at either A, B, or C_i, produces a 1 at the sum output, with no carry output. Row 7 produces the same output as procured from either row 4 or 5, because with any 2 inputs, the result is a 0 sum with a carry digit output. For row 8 we obtain both a sum output and a carry output. The A and B inputs alone would produce a 0 sum with 1 to carry, thus providing a C_o output. When, however, the 0 sum is added to the input carry digit, we have $0 + 1 = 1$ for the sum, thus providing for the dual output from the adder.

This table yields a variety of Boolean algebraic expressions. In

83

row 6, for instance, it is indicated that a sum output is produced for an A input, but not for B or C_i. Hence, the logic expression for row 6 is $A \cdot \bar{B} \cdot \bar{C}_i$. Similarly, row 8 shows that a sum output is obtained for inputs A, B, and C_i, or $A \cdot B \cdot C_i$. Row 7 has a carry output for $A \cdot \bar{B} \cdot C_i$. When we obtain the logic expression for a row which produces a logic 1 *sum output*, we obtain an expression of a *logic sum of the logic products of the input variables.* In such an instance it is called a *minterm* expression. (See the discussions on *sum of products* and *product of sums* in Chap. 4, Laws of Identity.)

If we write expressions for rows in which 0 appears at the output, we obtain expressions of the *logic product of the logic sums,* and this is called a *maxterm* expression. As will be discussed more fully in the next chapter, most of the algebraic expressions derived from truth tables contain redundancies. Methods for reducing redundant terms and simplifying expressions have already been covered in preliminary form in earlier chapters, and will be covered in greater detail in the chapters that follow.

GRAY CODE

The binary code discussed earlier in this chapter is also known as a *weighted code* since each digit has a value (weight) such as 16, 8, 4, 2, 1. By adding the weights of the digits indicated, the base-10 value of the group is ascertained: $1001 = 8 + 1 = 9$, etc. Special codes are also useful, however, and these are derived from the binary but are of the *unweighted* type, where the sum of the digit values (weights) is not equal to the number represented by the group of digits. One such unweighted code is the *Gray* code, also known as the *reflected binary* or *cyclic* code. Because it minimizes errors in certain applications, it is also termed a *minimum-error* code.

The Gray code is employed where physical changes such as shaft rotations, lever movements, or displacements must be converted to a binary-value equivalent, so that a digital computer can process the data for purposes of computation, control, or information storage. Hence the process finds application in guidance and tracking systems in military applications, as well as in industry (control of machine processes).

Sensing devices that convert physical changes to binary representations may introduce errors when several digits change in going from one binary number to another. Such is the case with binary, where, for instance, four numerals change when going from 0111 to 1000 (7 to 8), or from 1111 to 10000 (15 to 16), etc. The Gray code was designed so that only one numeral changes when going from one number to the next higher number, as shown in the following table.

Base-ten	Binary	Gray
0	0000	0000
1	0001	0001
2	0010	0011
3	0011	0010
4	0100	0110
5	0101	0111
6	0110	0101
7	0111	0100
8	1000	1100
9	1001	1101
10	1010	1111
11	1011	1110
12	1100	1010
13	1101	1011
14	1110	1001
15	1111	1000
	(etc.)	

Dr. Frank Gray, who originated this code, used specific principles in establishing the sequence of numbers making up the Gray code. To convert a binary number to its Gray-code equivalent, the binary number is added to itself without carrying, but the added number is first indexed (moved over) to the right for one place and the digit that extends beyond the original number is dropped. Thus, to convert the binary number 101 (5) to its Gray-code equivalent, the process is:

$$
\begin{array}{r}
101\, \\
+\ 10\, \\
\hline
111\,
\end{array}
\begin{array}{l}
(5) \\
1 \quad \text{(indexed to right)} \\
\text{(addition yields Gray number)}
\end{array}
$$

When a carry function is encountered it is ignored. The following indicates the process where the binary number 110 is converted to its Gray-code equivalent:

(binary) 110 (6)

  + 11 (indexed and extended numeral dropped)

(Gray) 101 (addition without carry)

Note that the addition of $1 + 1$ for the second-place numerals would normally equal 0 with 1 to carry. The carry function, however, was not performed. This principle holds true even though numerous digits are involved:

(binary) 1111 (15)

  + 111 (indexed and numeral dropped)

(Gray) 1000 (addition without carry)

Half adders (EXCLUSIVE OR circuits) can be used for converting the Gray code to binary for processing by a computer. The arrangement for five inputs is shown in Fig. 5-4. Because of the half-adder function, addition is performed without carry. Assume, for instance,

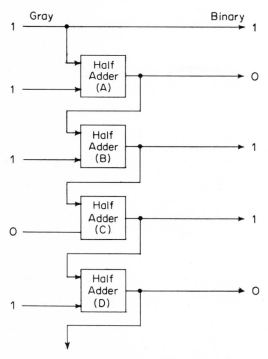

Fig. 5-4. Gray code to binary code conversion.

that the Gray-code number 11101 is to be converted. This number is applied to the inputs as shown, and the upper number 1 appears at the output as the leftmost digit of the binary number equivalent. The next numeral 1 enters the half adder (A), but because $1 + 1$ yields 0 in this instance, half adder (B) performs $1 + 0$ and a "1" output is produced. This is applied to half adder (C) and again, since $1 + 0$ is added, the output is 1. This numeral is applied to half adder D, which also receives a "1" input, for $1 + 1 = 0$ (with no carry). Thus, the output is the binary equivalent 10110.

The Gray code is also related to the truth maps of the multivariable variety, as discussed in Chap. 7.

ERROR-DETECTING CODES

The Gray code is a *minimum-error* code and must not be confused with *error-detecting* codes. A well-known code for detecting errors in computing or switching equipment is known as *parity check*. Parity is "the quality of being equal" and hence a parity check can be made by using a code that forms either an even number of digits or an odd number by adding a digit to the grouping. The additional digit is referred to as a *parity digit* and may be either a 1 or a 0. The parity digit is selected so that the number of all digits in the binary group is either even or odd. If the system is selected where *odd parity* prevails, for instance, the binary number plus the parity digit would have an odd number of "1's" in the group, as shown below:

Base-ten Number	Parity Digit	Binary Number
1	0	0001
2	0	0010
3	1	0011
4	0	0100
5	1	0101
6	1	0110

Since the parity digit is carried along with the binary number, an error is immediately detected because it would change the parity of the number to even.

If *even parity* is used, the parity digit is chosen so that the binary number plus the parity digit will have an even number of "1's" in the group, as shown below:

Base-ten Number	Parity Digit	Binary Number
1	1	0001
2	1	0010
3	0	0011
4	1	0100
5	0	0101
6	0	0110
	(etc.)	

COMPLEMENT NUMBERS

The power of the complementing process is well exemplified in computer applications, where it is used to simplify subtraction and to manipulate negative numerical values. Subtraction in binary arithmetic is fairly simple if no *borrowing* is involved, as shown by the following examples.

$$
\begin{array}{r}
1011\ (11) \\
-1001\ \ (9) \\
\hline
=0010\ \ (2)
\end{array}
\qquad
\begin{array}{r}
10111\ (23) \\
-\quad 101\ \ (5) \\
\hline
=10010\ (18)
\end{array}
$$

As in ordinary arithmetic, $1 - 1 = 0$ and the remaining binary number is the answer. The following examples involve the borrowing principle as in conventional arithmetic, except we must borrow from 10 (value 2) or from 100 (value 4), etc.

$$
\begin{array}{r}
101\ (5) \\
-\ 10\ (2) \\
\hline
11\ (3)
\end{array}
\qquad
\begin{array}{r}
100101\ (37) \\
-\quad 1001\ \ (9) \\
\hline
=\ 11100\ (28)
\end{array}
$$

In computer design it is often more expedient to dispense with the subtraction method, which involves the borrowing of numbers, and utilize a system whereby the subtraction is performed by complementing and addition procedures. Thus, computing processes are simplified because both addition and subtraction are performed by common circuitry.

To perform subtraction by addition, the subtrahend is converted to its complement and is added to the minuend. The leftmost bit is transferred to first place (*end-around carry*) and added to the remainder. (As explained in Chap. 4, we complement a number by changing each 0 to 1, and each 1 to 0.)

The following shows the complete process, first by ordinary subtraction and next by the complement-addition method:

$$10110 \text{ (minuend)}$$
$$- \ 1010 \text{ (subtrahend)}$$
$$\overline{\ 1100} \text{ (remainder)}$$

$$10110 \text{ (minuend)}$$
$$+ \ 10101 \text{ (complement of subtrahend)}$$
$$\overline{101011}$$
$$+ \qquad 1 \text{ (end-around carry of leftmost bit)}$$
$$\overline{\ 1100} \text{ (remainder)}$$

In using the one's complement process with end-around carry, the subtrahend must have as many binary bits (1 and 0) as are present in the minuend. Otherwise correct results will not always be obtained. Assume, for instance, that the following subtraction is to be processed by addition:

$$11110 \ (30)$$
$$- \ 1011 \ (11)$$
$$\overline{10011} \ (19)$$

If we do not fill out the subtrahend to 01011 we obtain a complement of 0100 which will not yield the right answer:

$$11110$$
$$+ \ \ 0100 \text{ (complement of 1011)}$$
$$\overline{100010}$$
$$+ \qquad 1 \text{ (end-around carry)}$$
$$\overline{11} \text{ (false remainder)}$$

By filling out the subtrahend to equal the number of bits in the minuend, we obtain the correct remainder:

$$11110 \ (30) \qquad 11110$$
$$-01011 \ (11) \qquad + \ 10100 \text{ (complement of 01011)}$$
$$\overline{10011} \ (19) \qquad \overline{110010}$$
$$\qquad\qquad\qquad + \qquad 1 \text{ (end-around carry)}$$
$$\qquad\qquad\qquad \overline{10011} \text{ (true remainder)}$$

The complement principle permits representation of negative arithmetical values in computers for internal processes using binary numbers. Thus, the binary positive number 1001 (9) becomes 0110 when it is a binary −9. When binary negative numbers are added to positive ones, the end-around carry process is used:

$$7 + (-3) = 4$$

$$
\begin{array}{l}
111 \text{ (7)} \\
+\ 100 \text{ (complement of 011)} \\
\hline
1011 \\
+\quad 1 \text{ (end-around carry)} \\
\hline
100 \text{ (= +4)}
\end{array}
$$

When negative numbers are added to each other, both complemented numbers are added and end-around carry again utilized:

$$(-3) + (-3) = -6$$

$$
\begin{array}{l}
100 \text{ (complement of 011)} \\
+\ 100 \\
\hline
1000 \\
+\quad 1 \text{ (end-around carry)} \\
\hline
001\ =\ -6 \text{ (complement of +6, 110)}
\end{array}
$$

In computers the negative values are identified by using an additional leftmost bit (sign bit) to indicate a negative value by a logic 1 representation for the sign bit.

Just as the complemented subtrahend in subtraction had to have as many bits as the minuend, bits magnitude is important in negative representations. For addition of negative values the magnitude of the sum determines the number of bits in the addend and the augend. Thus, if −8 is to be added to −9, the five-bit complements of −8 and −9 must be used:

$$
\begin{array}{l}
-8\ =\ 10111 \text{ (complement of 01000)} \\
-9\ =\ 10110 \text{ (complement of 01001)} \\
\hline
101101 \\
+\qquad 1 \text{ (end-around carry)} \\
\hline
01110\ =\ -17 \text{ (the complement of +17, 10001)}
\end{array}
$$

Addition of the type discussed is carried out conveniently in computers by operating in the parallel mode as more fully discussed in

Chap. 8. NOT circuits (inverters) may be used for changing binary numbers to their complements. A system for selecting either the original binary number or its complement is shown in Fig. 5-5. The literals

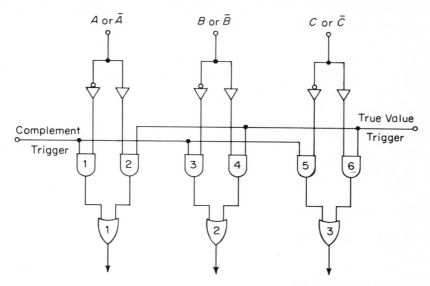

Fig. 5-5. Complementing circuitry.

A, B, C, etc. indicate individual entries into the system and it must be remembered that $A = 1$, $\bar{A} = 0$, $B = 1$, $\bar{B} = 0$, etc.

Note that if an A value is entered at the top, it appears as $\bar{A}$ at one AND circuit and as A at the other. Neither can enter the individual AND circuits until coincidence appears, because we must have $A \cdot A$ to obtain an output. If we want an A output we also apply a logic 1 to the true value trigger which applies this binary bit to successive inputs of the individual AND circuits.

Coincidence occurs only for AND gate 2, however, and the output from this circuit is applied to the first OR circuit for the final output from this system. Coincidence did not occur for the first AND circuit because the NOT circuit applied 0 to one input. For AND circuits 3, 4, 5, and 6, only the trigger input is applied to one input of each, hence no coincidence occurs.

If we had applied A and wanted $\bar{A}$ output, a logic 1 would have been applied to the complement trigger instead of the true-value trigger. At the first AND circuit the inputs would now be 1 *and* 0,

91

and since no coincidence occurs, we obtain a logic 0 output which is the complement of 1, or $\bar{A}$.

If $\bar{A}$ is entered originally to the system ($\bar{A} = 0$), the NOT inverts it and applies a logic 1 to the first AND circuit. Thus, if an A output (complement of $\bar{A}$) is required, the complement trigger would apply a coinciding logic 1 to the first AND circuit for obtaining an A output.

If A, B, and C are entered (111) we can obtain the true sum or the complement. The inputs could be mixed, such as $\bar{A}$, B, and $\bar{C}$; or A, $\bar{B}$, and $\bar{C}$, etc. and for complements we would obtain outputs of A, $\bar{B}$, and C; or $\bar{A}$, B, and C for the second group mentioned. (For the $\bar{A}$, B, and $\bar{C}$ inputs we would apply 010 and obtain 101 for the complement.)

Questions and Problems

1. Define the word *radix*.

2. What are the base-ten values for the binary numbers 10001, 10100, 10111, and 100001?

3. Write the binary numbers for 21, 22, 25, 26, and 27.

4. What binary sum is obtained when the binary number 01111 is added to 10100?

5. What binary sum is obtained when the binary numbers 101, 110, 001, and 111 are added?

6. Of what value is a half adder other than combining it with another half adder to form a full adder?

7. What must be included to combine half adders to form a full adder?

8. Define the words *minterm* and *maxterm*.

9. What is the difference between a *weighted code* and an *unweighted* one?

10. What is the advantage of the Gray code over the basic binary code?

11. Give two other names for the Gray code.

12. Write the Gray-code equivalents of the binary numbers 10001, 11011, 10111, and 11101.

13. What system may be used to convert a Gray-code number to its equivalent binary value?

14. Briefly explain the difference between a *minimum-error* code and an *error-detecting* code.

15. Which of the following numbers represent odd parity?

(a) 1 1011   (b) 1 0001

(c) 0 1101   (d) 1 1001

(e) 0 1110   (f) 1 1111

16. Redo the following subtraction process, using the 1's complement end-around carry principle:

$$\begin{array}{r} 11000\ (24) \\ -\ \ 1010\ (10) \\ \hline 1110\ (14) \end{array}$$

17. Subtract 5 from 21 by the conventional binary method and also show the equivalent 1's complement, end-around carry method.

18. What is the advantage of the complement method of subtraction by addition over the conventional subtraction method? Explain briefly.

19. Perform the following addition, using the 1's complement, end-around carry method: $16 + (-6)$.

20. Briefly define the term "sign bit."

6

PROCESSING BOOLEAN EXPRESSIONS

Combinational Circuits

INTRODUCTION

There are two basic logic-circuit types: the *combinational* and the *sequential*. In a combinational switching circuit the output is determined only by the existing input combinations. In a sequential switching circuit the output is determined not only by the existing inputs, but also by the history of previous inputs, as indicated by data-storage (memory) devices. Thus, in the sequential circuits the output expressions are not indicative of the Boolean logic functions of the existing inputs. A complete logic system may contain both combinational and sequential circuitry. In the first eight chapters of this text the combinational circuits are covered and the applied Boolean algebra discussed. Sequential circuitry starts with Chap. 9.

Simplification processes in switching systems are varied and some procedures are more suitable to certain logic expressions than to others. Proficiency is related to one's knowledge of Boolean algebra principles and practice in the selection of the most appropriate method. Techniques used can consist of Boolean algebraic manipulations or the use of logic maps. Both the algebraic and graphical methods are useful and each offers some advantages not present in the other. Basic algebraic processes, including the elimination of redundancies and superfluous segments of expressions, are covered in this chapter. The graphical map method is explored in Chap. 7.

94

SIMPLIFICATION TECHNIQUES

In the simplification of Boolean algebra expressions we utilize the various principles and rules covered in Chap. 4. Some of these axioms and laws are expanded in this section and applications illustrated to serve as a guide for simplification procedures.

Redundant or superfluous expressions can often be eliminated by inspecting the sequence of logic terms and relating them to their switching functions. Consider, for instance, the expression $A\bar{B} + AB$, which states that *A and not B or A and B* are true statements. When this is factored (distributive law) we obtain $A(\bar{B} + B)$. If $B = 1$, then $B + \bar{B} = 1$, the $A(\bar{B} + B)$ is $A \cdot 1$ and hence the reduced form becomes an *A* only. Similarly, $(A + B)(A + \bar{B})$ becomes $A + (B\bar{B})$ which, in essense, states that *A* is true, *or B and not B*. The variable $\bar{B}$ is 0, and since $1 \cdot 0 = 0$, the expression actually is $A + (0)$ which is *A* alone.

Expressions such as $A + AB$ and $A(A + B)$ can also be reduced to a single *A* as explained in Chap. 4, in the section Axioms of Switching Algebra, where the laws of indempotency are discussed. Note, however, the difference between $A + AB$ and the expression $A + \bar{A}B$. The latter equals $(A + \bar{A})(A + B)$ and since $A + \bar{A} = 1$, we combine the 1 with the $(A + B)$ to obtain $1 \cdot (A + B)$ which equals $A + B$. Such simplification methods can also apply to expressions as:

$$A(\bar{A} + B) = A\bar{A} + AB$$

and the first member $(A\bar{A})$ is $A \cdot 0 = 0$. The second member (added to 0) provides AB as the reduced form of $A(\bar{A} + B)$.

Care must be taken when negated variables or completely negated terms are involved. The expression $\bar{A}\bar{B} + \bar{A}B + A\bar{B}$, for instance, produces $\bar{A}(\bar{B} + B) + A\bar{B}$ when the first two members are factored and added to the last member. However, $\bar{A}(\bar{B} + B)$ is equivalent to $\bar{A}(0 + 1)$ which equals $\bar{A}$. When this is combined with the last member we have $\bar{A} + A\bar{B}$. Note that the $A\bar{B}$ is not the same as $A\bar{A}$ or $B\bar{B}$, where $1 \cdot 0 = 0$. When we have $A\bar{A}$ we state that *A but not A* is true, while $A\bar{B}$ states that *A but not B* is true. The logic expression "*A but not A*" in itself is contradictory and forms the $1 \cdot 0$ logic which equals 0.

For the expression $(\bar{A} + A)(\bar{A} + \bar{B})$ the negated variables must again be treated carefully. The first expression $(\bar{A} + A)$ is equivalent

95

to 1 (when the variable A is designated as 1). Hence, the final form is $\overline{A} + \overline{B}$.

The distributive law, where we factor the X in an expression such as $XY + XZ$ to form $X(Y + Z)$, can be applied to more lengthy expressions also. Thus, the A in the following expression can be factored as shown:

$$AB + ACD + A\overline{EF} = A(B + CD + \overline{EF})$$

In the factored form we say that A and B, or A and CD, or A and $\overline{EF}$ is true, which contains the same logic as the original expression.

As mentioned in Chap. 5, such sums of products are *minterm* expressions, and the same rule applies to the dual form, the *maxterm*, which refers to products of sums:

$$(A + B)(A + C + D)(A + \overline{E} + \overline{F}) = A + B(C + D)(\overline{E} + \overline{F})$$

This principle can also be used to combine two common variables and factor them as a single function. This can be illustrated by the expression:

$$ABC + A\overline{B}C + AB\overline{C}D$$

In this expression, AC is common to both the first and second terms, hence it behaves essentially as a single AC function and can be treated as such when factoring. The first two members $(ABC + A\overline{B}C)$ form $AC(B + \overline{B})$ which is equivalent to $AC(1)$, or AC only. The simplified section is then combined with the third term to form:

$$AC + AB\overline{C}D$$

When this is factored, we obtain:

$$A(C + B\overline{C}D)$$

When the $B\overline{C}D$ term is considered with the C term, the $B\overline{C}D$ simplifies to BD:

$$A(C + BD) \quad \text{or} \quad AC + ABD$$

A good point to remember is that a repetition of a term in another larger term of the expression makes the latter redundant. If we say

$A + AB$, we are saying that A is true, *or A and B*. Since A alone is true, we do not need the AB.

This principle is illustrated in the following minterm expressions where the AB term appears in larger terms and can be eliminated:

$$AB + ABC + ABC(D + E) = AB$$
$$ABCD + AB + ABD(\bar{E} + \bar{F}) = AB$$

In the following minterm expressions, the ABC term is common to larger terms and hence the latter are superfluous:

$$ABC + ABCDEF + ABC(D + F) = ABC$$
$$ABC(\bar{D} + C) + ABCD + ABC = ABC$$

The rule applies to maxterm expressions also:

$$(A + B + \bar{C})(A + B + \bar{C} + D) = A + B + \bar{C}$$

If we have a variable in two or more terms of an expression and such variables are complements, the complemented variable in the larger term can be considered redundant and eliminated:

$$AB + A\bar{B}C = AB + AC$$

Note that the second member of the original expression is not eliminated entirely, but only the complemented variable.

For additional simplification the resultant can be factored:

$$AB + AC = A(B + C)$$

As an additional example, consider the following:

$$A\bar{B}C + A\bar{B}\bar{C}D = A\bar{B}C + A\bar{B}D = A\bar{B}(C + D)$$

The rule again applies to maxterm expressions, as shown in the following:

$$(A + B)(A + \bar{B} + C) = (A + B)(A + C)$$

In an expression such as $WX + \bar{W}Y + XYZ$ an X appears in the first term and a Y in the second. Hence, the reappearance of these variables in XYZ makes it redundant and the expression is equivalent

97

to $WX + \overline{W}Y$. Similar simplification can, of course, be applied to the dual (maxterm) expression:

$$(W + X)(\overline{W} + Y)(X + Y + Z) = (W + X)(\overline{W} + Y)$$

Combinations of the complemented variable and term reappearances can be used when necessary:

$$\overline{A}BC + \overline{A}B + \overline{A}\overline{B}D + ABE$$

Here the second term $\overline{A}B$ appears in the larger first term, hence the latter is redundant. In the third term the $\overline{B}$ in $\overline{A}\overline{B}D$ is a complement of the B in $\overline{A}B$ and hence the $\overline{B}$ can be eliminated in this term. Similarly, in the fourth term (ABE) the A is a complement of the $\overline{A}$ in the second term $\overline{A}B$. Thus, by removing the $\overline{A}BC$ term, as well as the complements $\overline{B}$ and A, we obtain as a final form:

$$\overline{A}B + \overline{A}D + BE$$

In the expression $\overline{X}Y + X\overline{Y}Z$ we have the smaller term also appearing in the larger, but both the X and Y are complemented in the larger. *When more than one variable is complemented in a single expression, no redundancy exists* and hence the expression cannot be reduced additionally. Similarly, $\overline{X}YZ + X\overline{Y}$ cannot be simplified by elimination of complements.

In the expression $XY + \overline{X}Z$ it is important to note that the $\overline{X}$ in the second term is a complement of the X in the first term, but since the entire first term (XY) is not included in the second term, no term repetition exists and no redundancy is present. In the following, however, simplification is possible:

$$XY + \overline{X}Z + YZW = XY + \overline{X}Z$$

This is possible because the first two terms include Y and Z, hence the last expression which also contains these variables, is redundant. The dual (maxterm) expression is treated similarly:

$$(X + Y)(\overline{X} + Z)(Y + Z + W) = (X + Y)(\overline{X} + Z)$$

After repetitions and complements have been eliminated, the resultant expression may require factoring to obtain the most simplified form:

$$ABC + AB + \overline{A}D + BDE + A\overline{B}E = AB + \overline{A}D + AE$$
$$= A(B + E) + \overline{A}D$$

In the original equation the larger first term is eliminated because of the repetition of the AB in the second term. The B and D in the second and third terms are repeated in the fourth term BDE; hence this term is redundant and can be eliminated. The fifth term contains $\bar{B}$ which is a complement of the B in the second term, hence the fifth term becomes AE. The final form, when factored, is $A(B + E) + \bar{A}B$ as shown.

Summary of Minimum Forms

S1 $A + A = A$
$\quad\;$ $1 + A = 1$

S2 $A + \bar{A} = 1$
$\quad\;$ $1 + \bar{A} = 1$
$\quad\;$ $1 + 0 = 1$
$\quad\;$ $A + 0 = A$

S3 $A \cdot A = A$
$\quad\;$ $1 \cdot A = A$
$\quad\;$ $1 \cdot 1 = 1$

S4 $A \cdot \bar{A} = 0$
$\quad\;$ $A \cdot 0 = 0$
$\quad\;$ $1 \cdot 0 = 0$

S5 $(A + B)(A + C) = A + BC$

S6 $AB + AC = A(B + C)$

S7 $A(\bar{A} + B) = AB$

S8 $A(A + B) = A$

S9 $A + (\bar{A}B) = A + B$

S10 $A + AB = A$

S11 $A\bar{B} + AB = A(\bar{B} + B) = A$

S12 $(A + B)(A + \bar{B}) = A + (B\bar{B}) = A$

S13 $(\bar{A} + A)(\bar{A} + \bar{B}) = \bar{A} + \bar{B}$

S14 $(\bar{A} \cdot \bar{B}) + (\bar{A} \cdot B) = \bar{A}$

S15 $(\bar{A} \cdot \bar{B}) + (A \cdot \bar{B}) = \bar{B}$

S16 $AB + A\bar{B}C = AB + AC$

S17 $(A + B)(A + \bar{B} + C) = (A + B)(A + C)$

S18 $(A + B)(\bar{A} + C)(B + C + D) = (A + B)(\bar{A} + C)$

S19 $AB + \bar{A}C + BCD = AB + \bar{A}C$

99

Complementation

$$A = \overline{A}$$
$$(+) = (\cdot)$$
$$1 = 0$$
$$\overline{\overline{A}} = \overline{A}$$
$$AB + \overline{C} = (\overline{A} + \overline{B}) \cdot C$$

Dualization

$$A + B = A \cdot B$$
$$A + (0 \cdot \overline{B}) = A(1 + \overline{B})$$
$$(A + B)(C + D) = AB + CD$$

Equivalency (De Morgan's Theorem)

$$\overline{AB} = \overline{A} + \overline{B}$$
$$\overline{A + B} = \overline{A} \cdot \overline{B}$$

Minterm and Maxterm Derivations

Heretofore we have analyzed various logic expressions and cited rules for simplification. Such expressions may be derived from existing switching networks as previously shown. Often, however, a new combinational switch design is required and we must derive the necessary expression for analysis and possible simplification. In the design of logical switching circuits we are concerned with two factors: the input combinations that are to prevail, and the required output logic to be obtained for each input. From such data an orderly process may be followed to obtain the desired logical expression.

Initially, a truth table is prepared for the circuits or problem statement. Such a table should list all possible input-output functions. Next, Boolean algebra expressions are obtained from the truth-table statements. Finally, the expressions are reduced to their minimum form and redundancies eliminated.

As an example, assume we need a switching circuit having A, B, C, inputs and we want an output for all input combinations except B and BC. The resultant truth table would be:

Row	Inputs			Output
	A	B	C	
1	0	0	0	0
2	0	1	0	0
3	0	1	1	0
4	0	0	1	1
5	1	0	0	1
6	1	0	1	1
7	1	1	0	1
8	1	1	1	1

As mentioned in Chap. 5, truth tables of this type yield a variety of Boolean expressions. When we obtain a sum of products expression such as $\bar{A}\bar{B}C + \bar{A}BC + AB\bar{C}$, etc., it is known as a *minterm*. The dual of the minterm is the *maxterm*, which refers to the product of sums $(\bar{A} + \bar{B} + C)(\bar{A} + B + C)(A + B + \bar{C})$ etc. The minterm expression is obtained from the 1's output of the truth table, and the maxterm from the 0's output. For the table shown, the minterm members make up the following expression:

$$(\bar{A}\cdot\bar{B}\cdot C) + (A\cdot\bar{B}\cdot\bar{C}) + (A\cdot\bar{B}\cdot C) + (A\cdot B\cdot\bar{C}) + (A\cdot B\cdot C)$$

Note that for the first 1 output (row 4) we have zeros for A and B. Hence the expression is C, *but not A and B*. Thus, the A and B are shown in negated form: $\bar{A}\cdot\bar{B}\cdot C$. For row 5, the A produces a 1 output, but not B and C, hence: $A\cdot\bar{B}\cdot\bar{C}$. Similarly, for row 6, the A and C produce a 1 output, but not the B, hence: $A\cdot\bar{B}\cdot C$. The same logic applies to the other minterm expressions in row 7 and 8.

The truth table shows only three maxterm expressions (rows 1, 2, and 3) for the zero outputs. When the maxterm expressions are obtained from the truth table, the 1 states are negated. Thus, the complete maxterm expression is:

$$(A + B + C)(A + \bar{B} + C)(A + \bar{B} + \bar{C})$$

Note that in row 2 the B is represented by 1, hence the expression is $A + \bar{B} + C$. For row 3, both the B and C are represented by 1's, hence the expression is $A + \bar{B} + \bar{C}$. Since the complete maxterm expression only contains three members as opposed to the five-member minterm expression, the simplification procedures for this switching

101

system would be less involved if we used the maxterm for Boolean algebra manipulations. To show that the same results are obtained, however, and to indicate the variety of methods that can be employed for obtaining the same end result, both the minterm and maxterm expressions will be analyzed.

The sequence of inputs in the truth table, or the sequential arrangement of the minterm and maxterm expression members can follow any order. This was stated in the commutative law (4-9) which stated that the order in which we add or multiply a series of variables does not alter the end results. Hence, we could have listed the minterm expression in reverse order:

$$(A \cdot B \cdot C) + (A \cdot B \cdot \overline{C}) + (A \cdot \overline{B} \cdot C) + (A \cdot \overline{B} \cdot \overline{C}) + (\overline{A} \cdot \overline{B} \cdot C)$$

Several methods of simplification can be used. In our Summary of Minimum Forms, page 99, we can use rule S11 for the first two members of the expression given above:

$$(ABC) + (AB\overline{C}) = AB(C + \overline{C}) = AB \cdot 1 = AB$$

The same rule applied to the third and fourth members of the expression produces:

$$A\overline{B}(C + \overline{C}) = A\overline{B} \cdot 1 = A\overline{B}$$

Our expression is now:

$$(AB) + (A\overline{B}) + (\overline{A} \cdot \overline{B} \cdot C)$$

The first two members of the reduced expression $(AB) + (A\overline{B}) = A$ (rule S11) to form:

$$A + (\overline{A} \cdot \overline{B} \cdot C)$$

Finally, by rule S9, this becomes $A + \overline{B}C$.

The same answer is obtained when these methods are applied to the *maxterm* expression $(A + B + C)(A + \overline{B} + C)(A + \overline{B} + \overline{C})$.

By rule S12 the first two terms can be reduced to $A + C$. The resultant full expression then becomes:

$$(A + C)(A + \overline{B} + \overline{C})$$

By rule S17 this can be simplified to $(A + C)(A + \overline{B})$. If we now factor this, we obtain the final answer which is the same as that obtained for the minterm: $A + (\bar{B}C)$.

SIMPLIFICATION VARIATIONS

According to rules S1 and S3, $A + A = A$ and $A \cdot A = A$, hence we can often simplify switching functions by using a given term several times as required in Boolean algebra manipulations. Thus, for the maxterm expression $(A + B + C)(A + \bar{B} + C)(A + \bar{B} + \bar{C})$ we can repeat the middle term and simplify the expression by the following procedure:

$$(A + \bar{B} + C)(A + B + C) = A + C \quad \text{(using S12)}$$
$$(A + \bar{B} + C)(A + \bar{B} + \bar{C}) = A + \bar{B}$$
$$= (A + C)(A + \bar{B})$$

Now we again have the minimum maxterm expression and by factoring we obtain the same answer as before:

$$A + (\bar{B}C).$$

This method also applies to the minterm processing. Again repeating the same expression term, we obtain:

$$A \cdot \bar{B} \cdot C + A \cdot B \cdot C = AC$$
$$A \cdot \bar{B} \cdot C + A \cdot B \cdot \bar{C} = A(\bar{B}C + B\bar{C})$$
$$A \cdot \bar{B} \cdot C + A \cdot \bar{B} \cdot \bar{C} = A\bar{B}$$
$$A \cdot \bar{B} \cdot C + \bar{A} \cdot \bar{B} \cdot C = \bar{B}C$$
$$= AC + A(\bar{B}C + B\bar{C}) + A\bar{B} + \bar{B}C$$

As before, this reduces to $A + \bar{B}C$.

As a final method, we can form a dual of the maxterm expression to obtain the minterm and revert the final simplification back to its maxterm equivalent:

$$(A \cdot B \cdot C) + (A \cdot \bar{B} \cdot C) + (A \cdot \bar{B} \cdot \bar{C}) \quad \text{(dual of maxterm)}$$
$$(A \cdot \bar{B} \cdot C) + (A \cdot B \cdot C) = AC$$
$$(A \cdot \bar{B} \cdot C) + (A \cdot \bar{B} \cdot \bar{C}) = A\bar{B}$$
$$= A(\bar{B} + C) \text{ the dual of which is } A + (\bar{B}C)$$

Any one of the terms making up the complete expression can, of course, be used for the repeated term and the same end result is obtained. For the maxterm, assume we had selected the $(A + B + C)$ term for the one to be used:

$$(A + B + C)(A + \overline{B} + C) = A + C$$
$$(A + B + C)(A + \overline{B} + \overline{C}) = A + (B + C)(\overline{B} + \overline{C})$$
$$= A + B\overline{C} + \overline{B}C$$

Combining the new terms thus obtained gives us $(A + C)(A + B\overline{C} + \overline{B}C) = A + C(B\overline{C} + \overline{B}C) = A + \overline{B}C$. Similarly, had we taken the term $(A + \overline{B} + \overline{C})$ as the repeated term, we would obtain:

$$(A + \overline{B} + \overline{C})(A + \overline{B} + C) = A + \overline{B}$$
$$(A + \overline{B} + \overline{C})(A + B + C) = A + B\overline{C} + \overline{B}C$$

The resultant again produces the correct simplification:

$$(A + \overline{B})(A + B\overline{C} + \overline{B}C) = A + \overline{B}C.$$

The expression $A + \overline{B}C$ obtained by all these methods indicates an OR circuit and an INHIBITOR as shown in Fig. 6-1. Again, by the

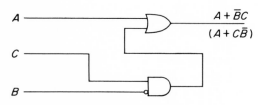

A

C

B

$A + \overline{B}C$

$(A + C\overline{B})$

Fig. 6-1. Logic diagram for $A + \overline{B}C$.

commutative law, $\overline{B}C = C\overline{B}$, hence the expression $A + C\overline{B}$ is also valid for this system.

As an additional illustration of the derivation of expressions from truth tables and the simplification of these expressions, the adder truth table shown on page 83 in Chap. 5 is repeated here, with the sum outputs and carry outputs grouped together for convenience in analysis:

	Inputs			Outputs	
Row	A	B	C_i	S	C_o
1	0	0	0	0	0
2	0	0	1	1	0
3	0	1	0	1	0
4	1	0	0	1	0
5	1	1	1	1	1
6	0	1	1	0	1
7	1	0	1	0	1
8	1	1	0	0	1

Note that for the sum (S) output there are four minterm and four maxterm expressions. For the minterm (which produces a 1 output) we have row 2 where it is C_i but not A or B; for row 3 it is B but not A or C_i; for row 4 it is A but not B or C_i; and for row 5 it is A, B, and C_i. Thus, the complete minterm expression for the sum (S) output for the three input variables A, B, and C, is:

$$S = \overline{A} \cdot \overline{B} \cdot C_i + \overline{A} \cdot B \cdot \overline{C}_i + A \cdot \overline{B} \cdot \overline{C}_i + A \cdot B \cdot C_i$$

This minterm expression is in minimum form and any attempts at simplification would only yield combinations such as $A(BC + \overline{B}\overline{C})$ without eliminating any letter variables.

$$\left. \begin{array}{l} A \cdot B \cdot C + \overline{A} \cdot \overline{B} \cdot C \\ A \cdot B \cdot C + \overline{A} \cdot B \cdot \overline{C} \\ A \cdot B \cdot C + A \cdot \overline{B} \cdot \overline{C} \end{array} \right\} \text{ not reducible}$$

(It would make no difference if we selected $\overline{A} \cdot \overline{B} \cdot C$ or $\overline{A} \cdot B \cdot \overline{C}$ for the repetitive term, because the expression can not be reduced an additional amount.)

For the adder truth table there are also four minterm rows producing a carry output (C_o). For row 5 it is A, B, and C_i; for row 6 it is B and C_i but not A, etc. Thus, the complete expression for the minterm carry output C_o is:

$$C_o = ABC_i + \overline{A}BC_i + A\overline{B}C_i + ABC_i$$

This expression lends itself to simplification by applying the rules previously discussed and illustrated. If we repeat the first member

105

and form pairs, we obtain the individual members of the simplified expression:

$$ABC + \bar{A}BC = BC(A + \bar{A}) = BC \cdot 1 = BC$$
$$ABC + A\bar{B}C = AC(B + \bar{B}) = AC \cdot 1 = AC$$
$$ABC + AB\bar{C} = AB(C + \bar{C}) = AB \cdot 1 = AB$$

Now when we combine the resultants of the foregoing simplification procedures, we obtain the minimum sum of products:

$$C_o = BC + AC + AB$$

We can now factor this [distributive law (4-11)] to obtain the product of sums form:

$$BC + A(C + B)$$

If we convert the final minimum sum of products (minterm) expression to its dual maxterm expression we obtain:

$$C_o = (A + B)(A + C)(B + C)$$

By the distributive law we can combine the first two terms to form $A + BC$. Adding this to the third member we have:

$$[A + (BC)] \cdot (B + C)$$

Additional reduction produces:

$$A(B + C) + (BC)$$

This is identical in logic to the expression $BC + A(C + B)$ obtained when we factored the final minterm expression $BC + AC + AB$. If, however, we had observed the same sequence of members, we would have obtained the dual as:

$$(B + C)(A + C)(A + B)$$

This would have produced $(C + AB)(A + B)$ and the final simplification is:

$$C(A + B) + AB$$

Here, again, though the expression appears to be different, we have actually realized identical logic to $BC + A(C + B)$, since we are stating that C *and* A is true, or C *and* B, or A *and* B, is true.

PROCEDURAL EXAMPLES

Example 1. Assume we need a switching system with three inputs, *A*, *B*, and *C*, where no single literal will produce an output, nor a *BC* combination alone. All other combinations produce an output. A truth table for these logic conditions is formed as follows:

A	*B*	*C*	Output
0	0	0	0
0	1	1	0
0	0	1	0
0	1	0	0
1	0	0	0
1	0	1	1
1	1	0	1
1	1	1	1

Since there are only three minterm rows, but five maxterm rows, the minterm expression would expedite simplification:

$$ABC + A\bar{B}C + AB\bar{C}$$

$$= AC(B + \bar{B}) \quad + AB\bar{C}$$

$$= AC \cdot 1 \quad\quad + AB\bar{C}$$

$$= AC + AB\bar{C} \ = AC + AB = A(B + C)$$

or:

$$ABC + A\bar{B}C = AC$$

$$ABC + AB\bar{C} = AB$$

$$= AC + AB = A(B + C)$$

The maxterm would, of course, yield the same answer. The expression, however, is lengthy:

$$(A + B + C)(A + \bar{B} + \bar{C})(A + B + \bar{C})(A + \bar{B} + C)(\bar{A} + B + C)$$
$$= A \cdot (A + B) \cdot (A + C) \cdot (B + C)$$
$$= A(B + C) \quad \text{[since } A(A + B) = A, \text{ etc.]}$$

The simplified minterm expression $AC + AB$ would require two

AND circuits feeding an OR circuit, as shown earlier in Fig. 4-8. With the factored expression (distributive law) we require only an OR and an AND circuit, as also shown in Fig. 4-8.

Example 2. A switching design with A, B, C, inputs requires that no output prevails for either B or C alone. For all other input combinations (or A alone) we want an output. The truth table would, therefore, be as follows:

A	B	C	Output
0	0	0	0
0	1	0	0
0	0	1	0
1	0	0	1
1	1	0	1
0	1	1	1
1	0	1	1
1	1	1	1

Because there are only three maxterm conditions this expression requires less steps for simplification:

$$(A + B + C)(A + \bar{B} + C)(A + B + \bar{C})$$
$$(A + B + C)(A + \bar{B} + C) = A + C$$
$$(A + B + C)(A + B + \bar{C}) = A + B$$
$$= (A + C)(A + B) = A + (BC)$$

This forms the logic circuit shown in the bottom portion of Fig. 4-8 in Chap. 4.

For comparison purposes, the minterm solution is as follows:

$$(A \cdot B \cdot C) + (A \cdot \bar{B} \cdot \bar{C}) + (A \cdot B \cdot \bar{C}) + (\bar{A} \cdot B \cdot C) + (A \cdot \bar{B} \cdot C)$$
$$A \cdot B \cdot C + A \cdot \bar{B} \cdot \bar{C} = A(BC + \bar{B}\bar{C})$$
$$A \cdot B \cdot C + A \cdot B \cdot \bar{C} = AB$$
$$A \cdot B \cdot C + \bar{A} \cdot B \cdot C = BC$$
$$A \cdot B \cdot C + A \cdot \bar{B} \cdot C = AC$$
$$= A(BC + \bar{B}\bar{C}) + AB + BC + AC$$
$$= A + BC$$

Example 3. As shown earlier in Fig. 6-1, *inhibiting* functions will appear as a NOT A or NOT B ($\overline{A}$ or $\overline{B}$) when logic circuitry involves this condition. As another example, assume that a switching system is required where an output is desired for A, B, or both A and B, but not when C is present. This produces the following truth table:

A	B	C	Output
0	0	0	0
0	1	0	1
1	0	0	1
1	1	0	1
0	0	1	0
1	0	1	0
0	1	1	0
1	1	1	0

Since there are three minterms, we have:

$$(A \cdot B \cdot \overline{C}) + (\overline{A} \cdot B \cdot \overline{C}) + (A \cdot \overline{B} \cdot \overline{C})$$

$$(A \cdot B \cdot \overline{C}) + (\overline{A} \cdot B \cdot \overline{C}) = B\overline{C}$$

$$(A \cdot B \cdot \overline{C}) + (A \cdot \overline{B} \cdot \overline{C}) = A\overline{C}$$

$$= B\overline{C} + A\overline{C}$$

$$= \overline{C}(B + A) \quad \text{or} \quad (B + A)\overline{C}$$

Thus our simplification indicates that we require an OR circuit and an AND circuit as shown in Fig. 6-2.

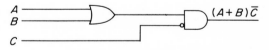

Fig. 6-2. Logic Diagram for $(A + B)\overline{C}$.

Example 4. When four literals such as A, B, C, D, are present, the greatly increased number of combinations may make it difficult to obtain a workable minimum sum or products (or product of sums) for simplification by algebraic manipulations. Methods detailed in subsequent chapters may be more applicable. When, however, there are only a few minterm or maxterm combinations, the four-literal members can be handled in similar fashion to the three-literal types. The following is an example where a few maxterm combinations are present:

A	B	C	D	Output
0	0	0	0	0
1	0	0	0	0
0	1	0	0	0

All other combinations produce 1 outputs.

Thus, we have the maxterm expression:

$$(A + B + C + D)(\bar{A} + B + C + D)(A + \bar{B} + C + D)$$

$$(A + B + C + D)(\bar{A} + B + C + D) = B + C + D$$

$$(A + B + C + D)(A + \bar{B} + C + D) = A + C + D$$

$$= (B + C + D)(A + C + D)$$

$$= AB + C + D$$

For the simplified expression an AND circuit and two OR circuits are indicated to yield AB or C or D. The logic diagram for this expression is shown in Fig. 6-3.

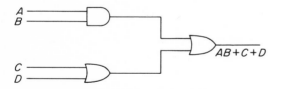

Fig. 6-3. Logic diagram for $(AB) + C + D$.

Example 5. On occasion the maxterm combinations may be numerous, but the minterm may not, for the four-literal logic, as shown by the truth table below.

A	B	C	D	Output
1	1	1	1	1
1	1	1	0	1
1	0	1	1	1

All other combinations produce 0 outputs.

Now the minterm expression is:

$$ABCD + ABC\bar{D} + A\bar{B}CD$$

$$ABCD + ABC\bar{D} = ABC$$

$$ABCD + A\bar{B}CD = ACD$$

$$= ABC + ACD$$

$$= AC(B + D)$$

This calls for an OR circuit and two AND circuits, as shown in Fig. 6-4.

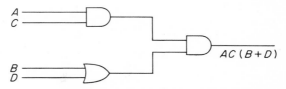

Fig. 6-4. Logic diagram for AC(B + D).

Example 6. The four variables may be involved in several levels of logic as shown by the following:

A	B	C	D	Output
0	0	0	0	0
0	0	1	0	0
0	1	0	0	0
0	0	0	1	0
0	1	0	1	0

All other combinations produce 1 outputs.

Here even the maxterm combinations become lengthy:

$$(A + B + C + D)(A + B + \bar{C} + D)(A + \bar{B} + C + D)$$
$$(A + B + C + \bar{D})(A + \bar{B} + C + \bar{D})$$

$$(A + B + C + D)(A + B + \bar{C} + D) = A + B + D$$
$$(A + B + C + D)(A + \bar{B} + C + D) = A + C + D$$
$$(A + B + C + D)(A + B + C + \bar{D}) = A + B + C$$

111

$$(A + B + C + D)(A + \bar{B} + C + \bar{D})$$
$$= A + (B + C + D)(\bar{B} + C + \bar{D})$$
$$= (A + C)(A + B + D)$$
$$= A + C(B + D)$$

Two OR circuits plus an AND circuit are needed as shown in Fig. 6-5, where A alone will produce an output, *or C and B or C and D*, or *C and both B and D* inputs.

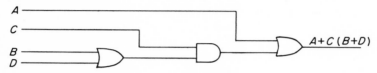

Fig. 6-5. Logic diagram for $A + C(B + D)$.

Example 7. Multi-input OR as well as AND circuits may be involved as shown by the following:

A	B	C	D	Output
1	1	1	1	1
1	1	1	0	1
0	0	1	1	1

All other combinations produce 0 outputs.
The complete minterm expression is:

$$(A \cdot B \cdot C \cdot D) + (A \cdot B \cdot C \cdot \bar{D}) + (\bar{A} \cdot \bar{B} \cdot C \cdot D)$$

$$(A \cdot B \cdot C \cdot D) + (A \cdot B \cdot C \cdot \bar{D}) = ABC$$

$$(A \cdot B \cdot C \cdot D) + (\bar{A} \cdot \bar{B} \cdot C \cdot D) = CD$$

$$= ABC + CD = C(D + AB)$$

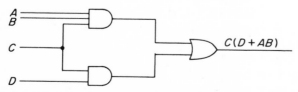

Fig. 6-6. Logic diagram for $C(D + AB)$.

As indicated by the final simplified expression, two AND circuits plus one OR circuit is involved, with one AND circuit having three inputs of ABC. The logic diagram is shown in Fig. 6-6.

Example 8. In Figs. 6-1 and 6-2 a single negated variable appeared in the simplified expression. On occasion, several negated literals may appear as shown by the following example:

A	B	C	D	Output
1	1	0	0	1
0	0	1	1	1
0	0	0	0	1

All other combinations produce 0 outputs.

The complete minterm expression is:

$$(\bar{A} \cdot \bar{B} \cdot \bar{C} \cdot \bar{D}) + (A \cdot B \cdot \bar{C} \cdot \bar{D}) + (\bar{A} \cdot \bar{B} \cdot C \cdot D)$$

$$(\bar{A} \cdot \bar{B} \cdot \bar{C} \cdot \bar{D}) + (A \cdot B \cdot \bar{C} \cdot \bar{D}) = \bar{C} \cdot \bar{D}$$

$$(\bar{A} \cdot \bar{B} \cdot \bar{C} \cdot \bar{D}) + (\bar{A} \cdot \bar{B} \cdot C \cdot D) = \bar{A} \cdot \bar{B}(\bar{C} \cdot \bar{D} + C \cdot D)$$

$$= (\bar{C} \cdot \bar{D}) + (\bar{A} \cdot \bar{B})$$

Thus, two AND circuits plus one OR circuit are needed, but since the output expression contains all negated variables, inverter (NOT)

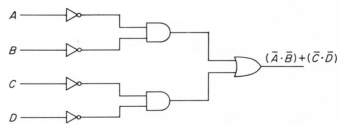

Fig. 6-7. Logic diagram for $(\bar{A} \cdot \bar{B}) + (\bar{C} \cdot \bar{D})$.

circuits are used as shown in Fig. 6-7. For an alternate circuit which satisfies this expression, see Fig. 3-8.

NOTE. While the objective is to realize the minimum circuitry for economical reasons, there may be occasions where other than the minimal form for a desired function may be preferred. In some instances specific inputs or outputs may be required, or some functions may already be present in associated circuitry that require a tie in with the system under question.

EXPANDED EXPRESSIONS

When minterm and maxterm expressions are obtained from a truth table, each term of the expression will, of course, have its full quota of literals. On occasion, however, we may have to simplify an expression in which one or more terms are incomplete. Assume, for instance, that we are given the expression:

$$(A + \bar{B})(A + B + C)$$

Here we must expand the shortened term $A + \bar{B}$ to include the missing literal C as well as its complement $\bar{C}$. Thus, the expanded version becomes:

$$(A + \bar{B} + C)(A + \bar{B} + \bar{C})$$

When we combine this with the second member of the expression we obtain:

$$(A + \bar{B} + C)(A + \bar{B} + \bar{C})(A + B + C)$$

Now we can use any of the various simplification processes to obtain the minimum form:

$$(A + \bar{B} + C)(A + B + C) = A + C$$
$$(A + \bar{B} + C)(A + \bar{B} + \bar{C}) = A + \bar{B}$$
$$= (A + C)(A + \bar{B})$$
$$= A + \bar{B}C$$

Minterms can, of course, also be expanded in this fashion. Consider, for instance, the expression

$$(X \cdot \bar{Y} \cdot \bar{Z}) + (Y \cdot \bar{Z})$$

The second member has a missing X, and hence must be expanded to include X and the complement $\bar{X}$:

$$X Y \bar{Z} + \bar{X} Y \bar{Z}$$

Combining the expansion with the remaining member of the original term we obtain:

$$(X \cdot \overline{Y} \cdot \overline{Z}) + (X \cdot Y \cdot \overline{Z}) + (\overline{X} \cdot Y \cdot \overline{Z})$$

The first two members now yield $X\overline{Z}$ and, using the second member with the last, yields $Y\overline{Z}$. Thus, $X\overline{Z} + Y\overline{Z} = \overline{Z}(X + Y)$ as the simplified result.

When a four-variable expression is used, two members may be missing, thus requiring four terms to complete the expansion. Assume we have

$$(AB\overline{C}D) + (AC)$$

From inspection, the second member has B and D missing, hence these two literals as well as their complements must be included to expand the term:

$$A\overline{B}CD + ABCD + ABC\overline{D} + A\overline{B}C\overline{D}$$

Now the complete expression becomes:

$$AB\overline{C}D + A\overline{B}CD + ABCD + ABC\overline{D} + A\overline{B}C\overline{D}$$

For simplification we can repeat the $ABCD$ member:

$$ABCD + AB\overline{C}D = ABD$$
$$ABCD + A\overline{B}CD = ACD$$
$$ABCD + ABC\overline{D} = ABC$$
$$ABCD + A\overline{B}C\overline{D} = AC(BD + \overline{B}\,\overline{D})$$
$$= AC + ABC + ACD + ABD$$

Since the smallest member is repeated in the next two, these cancel, leaving $AC + ABD$, or $A(C + BD)$.

In expanding a member of an expression we are not altering its basic logical value, since the addition of another literal plus its complement form $1 + 0 = 1$. Thus, the expansion of AB into three-variable terms adds the C and $\overline{C}$ to form:

$$ABC + AB\overline{C} = AB(C + \overline{C}) = AB(1 + 0) = AB$$

Thus, during the simplification process these two members will again

115

result in the AB expression. This is illustrated in the following example, using $(A \cdot B) + (A \cdot \bar{B} \cdot \bar{C})$.

Combining the expanded terms of AB with the second member of the expression we have:

$$(A \cdot B \cdot C) + (A \cdot B \cdot \bar{C}) + (A \cdot \bar{B} \cdot \bar{C})$$

Simplifying, we obtain:

$$(A \cdot B \cdot \bar{C}) + (A \cdot B \cdot C) = AB$$
$$(A \cdot B \cdot \bar{C}) + (A \cdot \bar{B} \cdot \bar{C}) = A\bar{C}$$
$$= AB + A\bar{C} = A(B + \bar{C})$$

For a maxterm the expanded versions involve $1 \cdot 0 = 0$. As an example, the following is the maxterm of the $(A \cdot B) + (A \cdot \bar{B} \cdot \bar{C})$ given above:

$$(A + B)(A + \bar{B} + \bar{C}) = (A + B + C)(A + B + \bar{C})(A + \bar{B} + \bar{C})$$

Note that $(A + B + C)(A + B + \bar{C})$ is equal to $A + B + (C \cdot \bar{C})$ which equals $A + B + (1 \cdot 0) = A + B + 1 = A + B$.

Here again the addition of the C and $\bar{C}$ cancels during the simplification process:

$$(A + B + \bar{C})(A + B + C) = A + B$$
$$(A + B + \bar{C})(A + \bar{B} + \bar{C}) = A + \bar{C}$$
$$= (A + B)(A + \bar{C}) = A + B\bar{C}$$

Converting this maxterm to its dual minterm we obtain $A(B + \bar{C})$.

Expressions may, of course, contain more than one abbreviated term as the following illustrates:

$$\bar{A}C + ABD$$

Here, the first term has B and D missing, hence four terms are required for expansion:

$$(\bar{A} \cdot B \cdot C \cdot D) + (\bar{A} \cdot \bar{B} \cdot C \cdot D) + (\bar{A} \cdot B \cdot C \cdot \bar{D}) + (\bar{A} \cdot \bar{B} \cdot C \cdot \bar{D})$$

The second term only requires the addition of C, hence only two terms are needed to expand:

$$ABCD + AB\bar{C}D$$

Thus, the complete expression is:

$$(\bar{A}\cdot B\cdot C\cdot D) + (A\cdot \bar{B}\cdot C\cdot D) + (\bar{A}\cdot B\cdot C\cdot \bar{D})$$
$$+ (\bar{A}\cdot \bar{B}\cdot C\cdot \bar{D}) + (A\cdot B\cdot C\cdot D) + (A\cdot B\cdot \bar{C}\cdot D)$$

Questions and Problems

1. **a.** Reduce, to its minimum form, the minterm expression:

$$A\bar{B} + A\bar{C}D + A\bar{E}F + AG\bar{H}$$

b. Change the following maxterm expression to its minterm dual and simplify:

$$(A + \bar{B})(A + \bar{C} + D)(A + \bar{E} + F)(A + G + \bar{H})$$

2. Simplify the expression:

$$(A + B + C)(A + \bar{B} + C)(A + B + \bar{C} + D) \quad = A + C(B+D)$$

3. **a.** Simplify the expression: $WX + WXY + WXY\bar{Z}(\bar{X} + W)$
   b. Reduce to a minimum the expression:

$$(W + \bar{X} + Y)(W + \bar{X} + Y + Z)$$

4. Eliminate redundancies in the following:
   a. $\bar{A}\bar{B} + ABC$
   b. $\bar{A}B\bar{C} + \bar{A}BCD$
   c. $(\bar{A} + \bar{B} + \bar{C})(\bar{A} + \bar{B} + C + D)$

5. Simplify the following expressions:
   a. $A\bar{B} + \bar{A}C + \bar{B}CD\bar{E}$
   b. $(A + \bar{B})(\bar{A} + C)(\bar{B} + C + D + \bar{E})$
   c. $\bar{A}BCD + \bar{A}B + ABE + (\bar{A}\cdot\bar{B}\cdot D)$
   d. $A\bar{B}(\bar{E}) + BD\bar{E} + \bar{A}D + AB + AB\bar{C}D(E + C)$

6. Reduce the following expressions:
   a. $(\bar{X}\cdot\bar{Y}) + (\bar{X}\cdot Y)$
   b. $(\bar{X}\cdot\bar{Y}) + (X\cdot\bar{Y})$
   c. $(\bar{X}\cdot\bar{Y}) + (WZ) + (\bar{X}Y)$

117

7. Complement the following expressions:
   a. $A\bar{B}C + D$
   b. $(A + C)(D + \bar{E})$
   c. $\overline{ABC} + \overline{DE}$

8. Form duals of the following expressions:
   a. $AB + (1 \cdot \bar{C})$
   b. $(\bar{A} + \bar{B})(C + D)(E + F)$
   c. $(A + B + D)(C + \bar{D})$

9. Define the words *minterm* and *maxterm* and explain why they are duals.

10. Simplify the following expressions:
    a. $(X \cdot \bar{Y} \cdot Z) + (X \cdot Y \cdot Z) + (X \cdot \bar{Y} \cdot \bar{Z})$
    b. $(\bar{X} + Y + Z)(X + \bar{Y} + Z)(X + \bar{Y} + \bar{Z})(X + Y + \bar{Z})(X + Y + Z)$

11. A switching system is needed having A, B, C, inputs. No outputs are to be present for zero input, or for BC input, or for C input only. All other combinations are to produce a 1 output. Write a truth table for these logic specifications and write the appropriate Boolean expression. Simplify the expression. Also, draw a diagram of the system indicated by the simplified expression.

12. Perform the requirements stated in Problem 11 for the following: A switching system is needed having A, B, C, inputs. An output is to be obtained only for AB, B alone, or for ABC. All other combinations produce a zero output.

13. From the following truth table portion, derive the minterm and solve for the simplified version. Draw a symbolic diagram for the answer obtained.

A	B	C	D	Output
1	1	1	1	1
1	1	0	0	1
0	0	1	1	1

All other combinations $= 0$.

14. Write the maxterm expression for the following truth table portion. Derive the minimum form, and draw a symbolic diagram:

W	X	Y	Z	Output
0	0	0	0	0
0	0	1	0	0
0	0	0	1	0

All other combinations = 1.

15. Write the minterm expression for the following truth table portion. Derive the minimum form, and draw a symbolic diagram:

A	B	C	D	Output
1	1	1	1	1
1	1	1	0	1
1	1	0	1	1

All other combinations = 0.

16. Write the minterm expression for the following truth table portion. Derive the minimum form, and draw a symbolic diagram:

A	B	C	D	Output
1	1	1	1	1
0	1	1	1	1
1	1	0	0	1

All other combinations = 0.

17. Obtain the Boolean expression from the following truth table, and reduce the expression to its minimum form. Prepare a symbolic diagram representative of the logic involved.

A	B	C	Output
0	0	0	0
1	0	0	1
0	0	1	1
1	1	0	0
0	1	1	0
1	0	1	1
0	1	0	0
1	1	1	0

119

18. Obtain the minimum sum or products for the following truth table, and draw the appropriate symbolic diagram for the minimum form:

A	B	C	Output
0	0	0	0
1	0	0	0
1	1	0	1
0	1	0	0
0	0	1	1
0	1	1	0
1	0	1	1
1	1	1	1

19. Write the maxterm expression for the following truth table portion. Derive the minimum form, and draw an appropriate symbolic diagram for it:

A	B	C	D	Output
0	0	0	0	0
1	0	0	0	0
0	1	0	0	0
0	0	1	0	0
0	1	1	0	0

20. Which of the following expressions is not in minimum form? Explain your answer.
    a. $AB + A\bar{B} + \bar{A}B$
    b. $XYZ + (\bar{X} \cdot Y \cdot \bar{Z}) + (X \cdot \bar{Y} \cdot \bar{Z}) + (\bar{X} \cdot \bar{Y} \cdot Z)$
    c. $(X \cdot Y \cdot Z) + (X \cdot Y \cdot \bar{Z}) + (X \cdot \bar{Y} \cdot Z) + (\bar{X} \cdot \bar{Y} \cdot Z)$

21. Expand the following expression, and obtain the simplified resultant:
$$A\bar{C} + ABC$$

22. Expand the following, and use the term $ABC\bar{D}$ of the full expression as the repeated member for simplification:
$$BC + A\bar{B}C\bar{D}$$

7

SIMPLIFICATION WITH
MULTIVARIABLE MAPS

Introduction

When a number of terms are present in a Boolean expression, algebraic manipulation often involves a lengthy and sometimes complicated process. Since there are 2^n functions of an n variable having two states (0 and 1), three such two-state variables can have 8 combinations ($2^3 = 8$), as shown in earlier chapters for truth tables involving A, B, C. Similarly, four two-state variables can have 16 combinations ($2^4 = 16$), and five such variables can have 32 combinations ($2^5 = 32$), etc. Thus, when the minterm and maxterm rows of a truth table are evenly divided, expressions can be lengthy and involved.

Logical maps provide for easy simplification of expressions, because the coincident algebraic functions overlap as the terms of the variables are entered. The reduced expression is then ascertained by visual inspection of the pattern configuration.

The maps and Venn diagrams introduced in Chap. 2 served to illustrate logic functions, but are seldom useful for simplification of multivariable expressions in *combinational* systems. For the *sequential* switching systems covered in Chap. 9 and Chap. 10, however, the maps discussed in Chap. 2 as well as the more complex types introduced in this chapter, are of considerable value.

For the combinational circuits, the Veitch-Karnaugh types are used

121

in which the square arrangements that correspond to variables conform to logical principles. Adjacent squares signify terms which differ by only a single variable, hence simplification of the resultant pattern is the end result. Some practice, however, is necessary in order to acquire facility in observational simplification.

TWO-VARIABLE MAP VARIATIONS

The square two-variable maps discussed in Chap. 2 can be modified to accommodate three variables as shown earlier in Fig. 2-6. The formation of the three-variable map is derived from the square map by converting the four represented values of the square map into a linear array as shown in Fig. 7-1. Note that the shaded areas representative

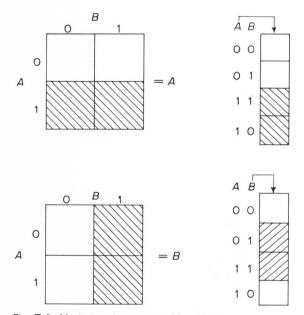

Fig. 7-1. Variations in two-variable maps.

of the variable A are at the bottom, adjacent to the 11 and 10 designations. For B, the two center squares are shaded, adjacent to the 01 and 11 numbers.

The adjacency of the two squares forming A (or B) is an important factor which will become more evident later in this chapter. Note, also, that the 0's and 1's representative of the variables A and B form the Gray code discussed in Chap. 5. Had the binary code been used, there would have been a separation of the shaded squares representative of the B variable:

Gray Code	Binary Code	
00	00	
01	01	Note separation of 1's
11	10	in B column
10	11	

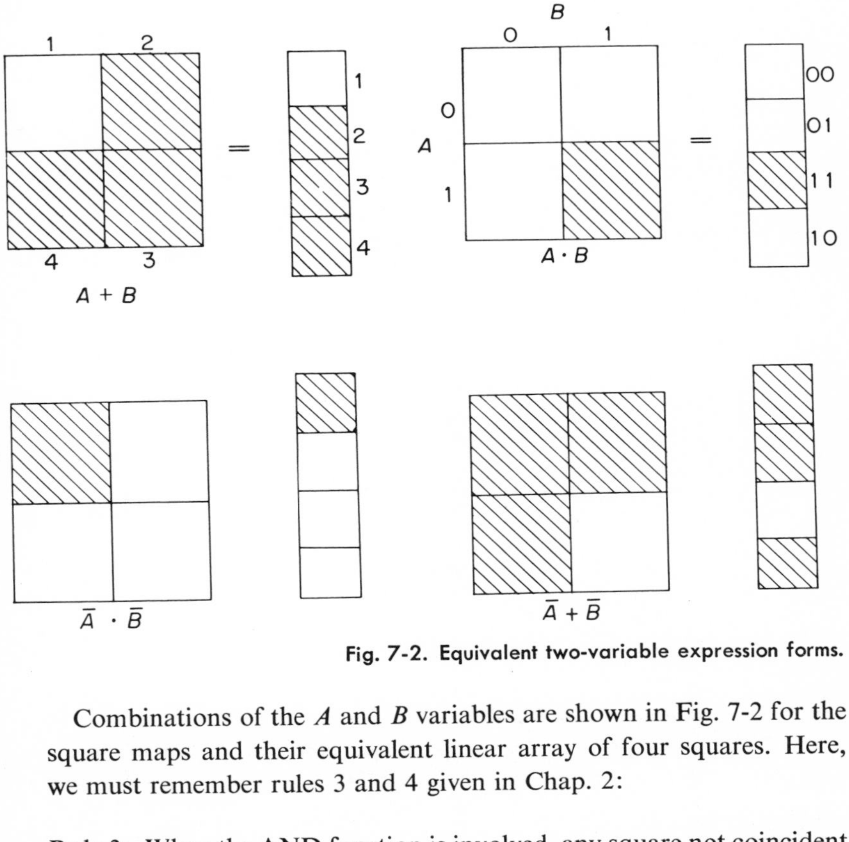

Fig. 7-2. Equivalent two-variable expression forms.

Combinations of the A and B variables are shown in Fig. 7-2 for the square maps and their equivalent linear array of four squares. Here, we must remember rules 3 and 4 given in Chap. 2:

Rule 3: When the AND function is involved, any square not coincident is left blank.

Rule 4: When the OR function is involved, all shaded squares remain intact.

For the $A + B$ expression at the upper left of Fig. 7-2, the squares are numbered to show the equivalent placement between the two map types. Because the OR function is involved, all shaded squares remain. For the AB function at the upper right, all squares not coinciding are blank. Negations of the A and B variables are shown in the lower portion of Fig. 7-2.

The single $\bar{B}$ variable for the square map and the linear-array equivalent is illustrated at the top of Fig. 7-3. Since it is important to keep

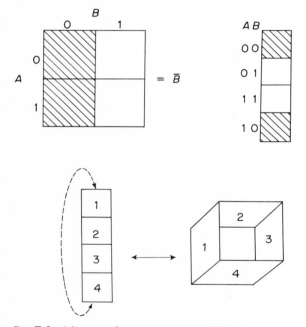

Fig. 7-3. Adjacency factors in two-variable maps.

the shaded squares adjacent, it would appear that the linear array is invalid because one shaded area is at the top and the other at the bottom. In the square map at the top left of Fig. 7-2, the squares identified by 1 and 4 are adjacent. Thus, for the linear array we must consider it as being bent around so that the squares 1 and 4 touch, as shown at the lower right.

124

THREE-VARIABLE MAPS

To accommodate three variables, an additional array is added to the two-variable type as shown in Fig. 7-4. To conform to the logic involved, four shaded areas are used to represent any single variable.

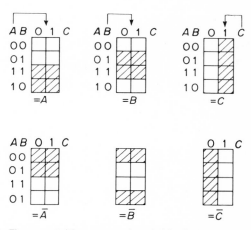

Fig. 7-4. Three-variable maps and variable designations.

For A, the squares opposite 11 and 10 are shaded, as shown at the upper left. Since $\bar{C} + C = 0 + 1 = 1$, this portion of the expression cancels and is omitted as part of the resultant logic. Hence, for the horizontal row identified by 11 we have only AB ($AB\bar{C} + ABC$). For row 10 we have $A\bar{B}$ with C again cancelled, $(A \cdot \bar{B} \cdot \bar{C}) + (A \cdot \bar{B} \cdot C)$. Combining the two resultants we have $AB + A\bar{B} = A$. (The *minterm*, *sum of products*, expression is used. The dual, maxterm, will be discussed later.) The applied logic need not be considered each time if we simply remember that *any two shaded horizontal rows cancel the C function*, leaving only the AB expressions designated by the Gray-code column.

For the B designation, note that the cancelled C blocks leave $\bar{A}B + AB = B$. For C, the entire vertical right column is shaded, and for $\bar{C}$, the left column is shaded as shown at the right of Fig. 7-4. The negated forms for A and B are shown at the lower left. For $\bar{B}$, the "wrap-around" adjacency factor discussed earlier for the two-variable map and shown in Fig. 7-3 also applies, and the lower and upper

125

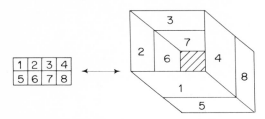

Fig. 7-5. Adjacency factors in three-variable maps.

shaded areas are considered adjacent. The adjacency factor for the three-variable map is illustrated in Fig. 7-5.

Two variable combinations for the three-variable maps are shown in Fig. 7-6. At the upper left, an OR function is indicated, hence both the A and C shaded areas are retained. For the A *and* C combination, however, only the two squares shown are coincidental and retained; all other A and C squares are omitted. For the $A + B$ at the upper right, all involved shaded areas are retained, while for $A \cdot B$ at the lower left, only those A and B squares which coincide are left intact. Similar considerations are involved with the BC and $B + C$ shown.

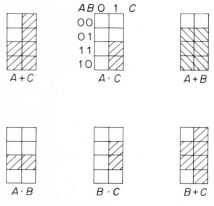

Fig. 7-6. Two-variable combinations in three-variable maps.

The same rules also apply to negated forms. For any AND function, one can write in the appropriate variable designations as A, or $\overline{A}$, B or $\overline{B}$, etc., and erase all those where no coincidence occurs as shown in Fig. 7-7. For the map at the upper left, for instance, the A occupies the lower four squares while the $\overline{C}$ is represented by the left vertical column. Since only two squares show two variables in each, these form the $A \cdot \overline{C}$ expression. Similar procedures can be undertaken for other expressions, such as the $B\overline{C}$, $\overline{B} \cdot \overline{C}$, and the $\overline{B}C$ illustrated.

Some additional examples of multivariable combinations are shown in Fig. 7-8. The upper left illustrates $\overline{A}B$. Since $\overline{A}$ alone would occupy the upper four squares and B the middle four, coincidence only occurs for the two shown. At the top center we have $(\overline{A}B) + C$. As men-

126

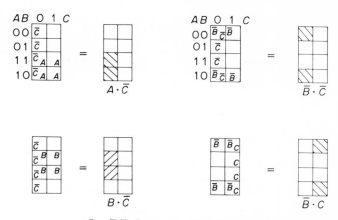

Fig. 7-7. Formation of AND maps with negations.

tioned earlier, when a horizontal row has both squares shaded we have $\bar{C} + C = 1$; hence for the second horizontal row shown the resultant is $\bar{A}B$. The vertical row is shaded to represent C, and since those squares which are *not* coincident are also shaded, an OR function is indicated, resulting in $\bar{A}B + C$.

For the $A \cdot \bar{B}$ shown at the upper right, the upper shaded blocks, which are part of $\bar{B}$, are not shown because no coincidence prevails. When this expression is combined with C we have the map illustrated at the lower left of Fig. 7-8. For understanding the bottom center map of $(AC) + (\bar{B}C)$ compare the AC shown in Fig. 7-6 with the $\bar{B}C$ shown in Fig. 7-7. For the lower right expression $(\bar{A} \cdot \bar{B}) + (\bar{B} \cdot \bar{C})$,

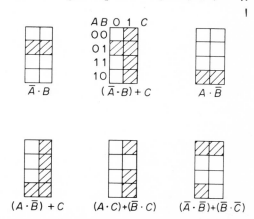

Fig. 7-8. Formation of multivariable combinations.

127

$\overline{A}\cdot\overline{B}$ only would leave the upper two squares shaded because they would be the only ones where $\overline{A}$ and $\overline{B}$ squares coincide. For $\overline{B}\cdot\overline{C}$ only, the upper left and lower left squares of this map would be shaded. Thus, the shading that is shown indicates the combined functions of $\overline{A}\cdot\overline{B}$ or $\overline{B}\cdot\overline{C}$.

Mapping and Simplification
of Three-variable Expressions

In mapping procedures the minterm expression is derived from a truth table and the three-variable expressions entered; overlapping where similar-value variables occur. Assume, for instance, that we are to map the expression $(\overline{A}\cdot B\cdot\overline{C}) + (A\cdot B\cdot\overline{C})$. For the map shown at the upper left of Fig. 7-9, the $\overline{A}B$ portion of $\overline{A}B\overline{C}$ is represented by the 01 (second from top) row of squares. Since $\overline{C}$ is also part of this expression, the 0 vertical C column is involved. Only the square next

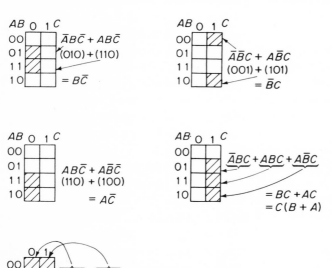

Fig. 7-9. Mapping three-variable expressions.

to the 01 is shaded, however, *to link in the adjacent* 01 *values.* Thus, the $\overline{A}B\overline{C}$ is represented by 010 binary equivalent. Similarly, for $AB\overline{C}$ we select the third-from-top row designated by 11 for AB and shade in the 0 square of the C column. Now we have represented $AB\overline{C}$ by 110.

The simplified expression $B\overline{C}$ is immediately in evidence. Since no two horizontal squares (side-by-side) are shaded, there is no cancellation of the $\overline{C}$ value. The two squares which are shaded are in a vertical plane representing $\overline{A}B$ as well as AB to give us $\overline{A}B + AB$ which equals B alone. Hence we have B and $\overline{C}$ as the simplified expression. The logic involved in deriving the vertical expressions need not be considered each time. If, in the vertical plane, two adjacent shaded squares have coinciding 1's, the appropriate variable is the true value. If the two adjacent shaded vertical squares have coinciding 0's in common, the appropriate variable designation heading the vertical column is the true value.

This adjacency factor is a vital point in undertaking simplification by visual inspection of shaded-area maps. Initially it is difficult to comprehend, but quite simple when the process is fully understood. As another example, consider the expression $(\overline{A} \cdot \overline{B} \cdot C) + (A \cdot \overline{B} \cdot C)$ mapped at the upper right of Fig. 7-9. The $\overline{A} \cdot \overline{B} \cdot C$ expression involves only one shaded square (the upper square of the logic-1 C vertical column). This represents 001 and hence $\overline{A} \cdot \overline{B} \cdot C$. Similarly, the $A\overline{B}C$ expression involves only the lower 1 square in the vertical C column for 101. Because of the wrap-around factor, however, the upper and lower squares are considered adjacent (Fig. 7-5). The upper and lower squares have two common 0's representing $\overline{B}$. This variable, with the C value produces the simplified expression $\overline{B}C$.

In the center left map of Fig. 7-9 the expression $AB\overline{C} + A\overline{B}\overline{C}$ again involves only two shaded squares which represent $\overline{C}$. The two vertical squares have two common 1's in the A column, hence the final expression is $A\overline{C}$. Note that when the expression $\overline{A}BC + ABC + A\overline{B}C$ is mapped, three squares are shaded in the vertical 1's column of the C variable. Thus we have a C variable initially, but now the center shaded square is adjacent to both the top shaded square as well as the lower shaded square. The center and top squares have 1's in common in the B column, hence we have BC. The center square and the bottom square have 1's in common in the A column, hence we also have AC. Thus, $BC + AC = C(B + A)$.

The lower left map of Fig. 7-9 involves the expression

$$(\overline{A} \cdot \overline{B} \cdot \overline{C}) + (\overline{A} \cdot \overline{B} \cdot C) + (\overline{A} \cdot B \cdot \overline{C}).$$

Note that the first term $\overline{A} \cdot \overline{B} \cdot \overline{C}$ would involve shading only the top 0 square of the vertical C column. The second term $\overline{A} \cdot \overline{B} \cdot C$, however, also has the $\overline{A} \cdot \overline{B}$ in common with the first term, hence the top horizontal row of the map is used again. Since the $\overline{A} \cdot \overline{B}$ portion is already represented, the 1's square of the C column is shaded to represent the C variable in the second term $\overline{A} \cdot \overline{B} \cdot C$. For the third term $\overline{A}B\overline{C}$, the 01 horizontal row is used, shading the 0 square of the C column for $\overline{C}$. This 010 horizontal representation thus stands for $\overline{A}B\overline{C}$.

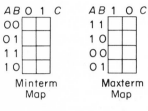

Minterm Map   Maxterm Map

Fig. 7-10. Minterm and maxterm maps.

Since the 0 and 1 squares of the top horizontal row are shaded, the $\overline{C} + C$ cancel, leaving $\overline{A} \cdot \overline{B}$. The upper left square, however, is common with the left lower square in the 01 horizontal row. Since the two 0's are common here under the A column, we obtain $\overline{A} \cdot \overline{C}$. Thus, we now have $(\overline{A} \cdot \overline{B}) + (\overline{A} \cdot \overline{C})$ which, when reduced by the distributive law, equals $\overline{A}(\overline{B} + \overline{C})$.

MINTERM AND MAXTERM MAPS

While the minterm expressions are all we need for simplification by the mapping process, we should also be familiar with the dual or maxterm version for an in-depth understanding of the logic involved. A maxterm dual of the minterm map appears as shown in Fig. 7-10, where the two are shown for comparison purposes.

Note that in the maxterm map all 1's and 0's are the reverse of those in the minterm map. This follows the logic previously discussed where the maxterm expressions use negative literals for the logic 1's, and positive literals for logic 0's. Thus, a truth table row such as 101 has a maxterm expression $\overline{A} + B + \overline{C}$. A comparison of how representative terms are mapped is given in Fig. 7-11.

To show how the same simplified term is obtained by using either the minterm or maxterm map, we will use the following truth table. This truth table was shown earlier in Chap. 6 for obtaining the minterm and maxterm expressions for purposes of illustration:

130

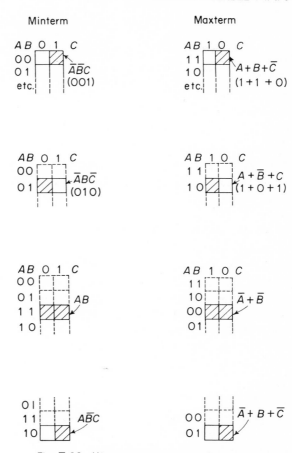

Fig. 7-11. Minterm-maxterm mapping comparisons.

A	B	C	Output	
0	0	0	0⎫	
0	1	0	0⎬ $(A + B + C)(A + \bar{B} + C)(A + \bar{B} + \bar{C})$	
0	1	1	0⎭	
0	0	1	1⎫	
1	0	0	1⎪	
1	0	1	1⎬ $(\bar{A} \cdot \bar{B} \cdot C) + (A \cdot \bar{B} \cdot \bar{C}) + (A \cdot \bar{B} \cdot C)$	
1	1	0	1⎪ $+ (A \cdot B \cdot \bar{C}) + (A \cdot B \cdot C)$	
1	1	1	1⎭	

The resultant minterm and maxterm maps are shown in Fig. 7-12. Note that the entry of the last four terms forms an equivalent single A

131

Minterm

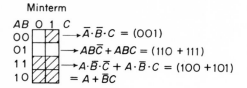

$$\overline{A}\cdot\overline{B}\cdot C = (001)$$
$$AB\overline{C} + ABC = (110 + 111)$$
$$A\cdot\overline{B}\cdot\overline{C} + A\cdot\overline{B}\cdot C = (100 + 101)$$
$$= A + \overline{B}C$$

Maxterm

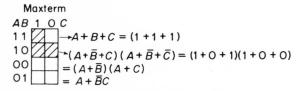

$$A + B + C = (1 + 1 + 1)$$
$$(A + \overline{B} + C)(A + \overline{B} + \overline{C}) = (1 + 0 + 1)(1 + 0 + 0)$$
$$= (A + \overline{B})(A + C)$$
$$= A + \overline{B}C$$

Fig. 7-12 Identical simplification from dual maps.

expression. Since no squares of this group are unshaded it represents an OR function. The upper single square formed from $\overline{A}\cdot\overline{B}\cdot C$ has adjacency with the lower right square, hence two 0's are common in the B column for $\overline{B}$. Since the latter are in the vertical 1's column of the C variable, we have $\overline{B}C$. When this is combined with the A we get $A + \overline{B}C$ as our final simplified expression.

In the maxterm map at the bottom of Fig. 7-12 only three shaded squares appear. For the two shaded horizontal squares the C value cancels, leaving $A + \overline{B}$. The two vertical shaded squares have 1's in common in the A column, hence equal $A + C$ (since the 1's squares of the C column are also involved). Thus we have: $(A + \overline{B})(A + C)$ which reduces to $A + \overline{B}C$ by the distributive law.

Note the duality which exists between the completed maps. For each shaded square in one map there is a corresponding unshaded square in the other. Thus, the unshaded squares in the minterm map form the shaded-square area in the maxterm map.

EXAMPLES AND ANALYSES

Example 1. The minterm rows from a truth table can, of course, be transferred directly to the map without converting them to the A, B, C, literals. Consider, for instance, the following truth-table portion:

132

A	B	C	Output
1	1	1	1
1	0	1	1
0	0	1	1

Other combinations produce 0.

Here, the first row 111 of the truth table calls for a shaded square in the 1's column of the C, opposite the 11 row of the map. Similarly, the 101 of the truth table calls for a shaded square in the 1's column of the C, opposite the 10 row of the map. The last row of the truth table 001 also involves only one shaded square in the 1's row of the C column of the truth table, opposite the 00, as shown at the top left of Fig. 7-13. The adjacent blocks opposite 11 and 10 have 1's in common

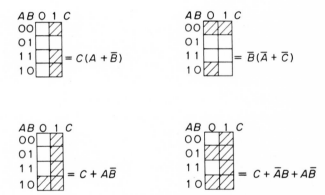

Fig. 7-13. Maps for Examples 1 to 4.

in the A column, and hence yield AC. The wrap-around factor links the bottom and top shaded areas where 0's are common in the B column, producing $\bar{B}C$. Thus, our complete expression is $AC + \bar{B}C$ which reduces to $C(A + \bar{B})$.

Example 2. The expression $(\bar{A} \cdot \bar{B} \cdot C) + (\bar{A} \cdot \bar{B} \cdot \bar{C}) + (A \cdot \bar{B} \cdot \bar{C})$ is mapped at the upper right of Fig. 7-13. Because the first two members of the expression are identical except for the C and $\bar{C}$, both squares are shaded in the map opposite the 00 row. The last member of the expression shades only the square in the 0's column for C opposite the 10 row. The two upper horizontal shaded squares provide an $\bar{A} \cdot \bar{B}$ term. The

133

wrap-around factor for the bottom and top $0 = C$ squares have two 0's in common under the B column, yielding $\bar{B} \cdot \bar{C}$. Hence, we have $(\bar{A} \cdot \bar{B}) + (\bar{B} \cdot \bar{C})$ which reduces to $\bar{B}(\bar{A} + \bar{C})$.

Example 3. If an expression has five minterm members and three maxterm, algebraic solutions would be directed toward the smaller maxterm expression. In mapping, however, the procedure is so simple that there is no need to resort to the maxterm maps. By confining the procedures to the minterm, greater familiarity and facility will be achieved.

Thus, a five-member minterm expression such as

$$(\bar{A} \cdot \bar{B} \cdot C) + (\bar{A} \cdot B \cdot C) + (A \cdot B \cdot C) + (A \cdot \bar{B} \cdot C) + (A \cdot \bar{B} \cdot \bar{C})$$

is easily mapped to produce the shaded areas shown at the lower left of Fig. 7-13. From inspection we have a totally shaded 1's column for C which is an OR function. The bottom adjacent squares cancel out the C value, leaving only $A\bar{B}$. Thus, the simplified expression is $C + A\bar{B}$.

Example 4. If a given expression contains some two-variable members as well as some three-variable ones, the two variable must be treated as having $0 + 1$ take the place of the missing member. Thus, if AB is given, it assumes that the full expression is $AB(\bar{C} + C)$ or $AB(0 + 1)$. This is the same as $AB\bar{C} + ABC$ discussed under *Expanded Expressions* on page 115 of Chap. 6. Thus, the two horizontal squares for $C = 0$ and $C = 1$ are shaded. Assume, for instance, that our expression is: $(\bar{A} \cdot \bar{B} \cdot C) + (\bar{A} \cdot B) + (A \cdot B \cdot C) + (A \cdot \bar{B})$. The resultant map is shown at the lower right of Fig. 7-13. For the $\bar{A} \cdot \bar{B} \cdot C$ member, the top 1's square of the C column is shaded. For the second member, $\bar{A}B$, we shade both squares opposite the 01 row as we would for $\bar{A}B\bar{C} + \bar{A}BC$. The third member ABC requires shading of the $1 = C$ square in the 11 row. For the last member, $A\bar{B}$, we again assume $(\bar{C} + C)$ as would occur in $(A \cdot \bar{B} \cdot \bar{C}) + (A \cdot \bar{B} \cdot C)$ and thus fill in both bottom squares opposite the 10 row. As shown, the entirely-shaded $1 = C$ column gives us an OR function for $C + \bar{A}B + A\bar{B}$.

Example 5. Another method for representing the three-variable map is to show it in a horizontal position and use 1's instead of shaded areas as shown in Fig. 7-14. At the top are two maps for the same expression,

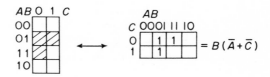

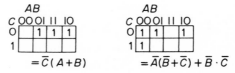

Fig. 7-14. Maps for Examples 5 to 7.

one showing the shaded procedure and the other the 1's designations. Because this type of map representation is often used, the following examples will use this method for familiarization.

In the two equivalent maps at the top of Fig. 7-14, the equation was $\overline{A}B\overline{C} + \overline{A}BC + AB\overline{C}$. Note that the horizontal map represents this expression in identical fashion to the shaded vertical type. The single shaded area opposite the 11 row for one map is duplicated in the other map by a 1 in the square corresponding to $0 = C$ adjacent to the 11 (AB) designations.

Example 6. The lower left map of Fig. 7-14 is for the expression: $(\overline{A} \cdot B \cdot \overline{C}) + (A \cdot B \cdot \overline{C}) + (A \cdot \overline{B} \cdot \overline{C})$. Because no adjacent 0 and 1 squares in the C columns are shaded, there is no C-value cancellation. Adjacent squares 11 and 10 have two 1's in common for the A value, hence we have $A\overline{C}$. The $0 = C$ square beside the 11 is also adjacent to the $0 = C$ square for 01. Hence, we have two 1's in common for B, yielding $B\overline{C}$. The complete expression is $A\overline{C} + B\overline{C}$ which reduces to $\overline{C}(A + B)$.

135

Example 7. Another instance of the adjacency of end squares is shown at the lower right of Fig. 7-14. This map is for the expression:

$$(\overline{A}\cdot\overline{B}\cdot\overline{C}) + (\overline{A}\cdot\overline{B}\cdot C) + (\overline{A}\cdot B\cdot\overline{C}) + (A\cdot\overline{B}\cdot\overline{C}).$$

The two adjacent squares in the 00 row give $\overline{A}\cdot\overline{B}$, and the adjacency of the 00 and 01 squares for $0 = C$ have 0's in common for A, producing $\overline{A}\cdot\overline{C}$. The wrap-around factor of the end squares have 0's in common for B, yielding $\overline{B}\cdot\overline{C}$. Hence, our total expression is

$$(\overline{A}\cdot\overline{B}) + (\overline{A}\cdot\overline{C}) + (\overline{B}\cdot\overline{C}).$$

which reduces to $\overline{A}(\overline{B} + \overline{C}) + (\overline{B}\cdot\overline{C})$.

Example 8. The first map in Fig. 7-15 is for the expression

$$ABC + AB\overline{C} + A\overline{B}C$$

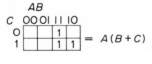

and is straightforward. The first two portions of the expression places 1's in the 11 (AB) row, thus cancelling the C value, leaving only AB. The last part, $A\overline{B}C$, places a 1 in the $1 = C$ square of the 10 row. The adjacent horizontal 1's produce AC, hence we have

$$AB + AC = A(B + C)$$

Fig. 7-15. Maps for Examples 8 and 9.

for the simplified resultant.

Example 9. The second map in Fig. 7-15 is for the expression: $(A\cdot B\cdot C) + (A\cdot\overline{B}\cdot\overline{C}) + (A\cdot B\cdot\overline{C}) + (A\cdot\overline{B}\cdot C) + (\overline{A}\cdot B\cdot C)$. The four right-hand squares with 1's gives us an A OR value. Adjacency occurs between the single 1 in the 01 row and the $1 = C$ in the 11 row. Since two 1's are present for the B value, we have BC. The simplified term is, therefore, $A + BC$.

FOUR-VARIABLE KARNAUGH MAPS

A rectangular map of 16 squares is used for the simplication of four-variable expressions, as shown in Fig. 7-16. The Gray-code notation is applied to both the vertical and horizontal columns as shown. The wrap-around principle applies to the top and bottom as well as to the sides as shown in Fig. 7-17. This dual adjacency can be pictured as ring formation as shown, where squares numbered 1 and 13 are adjacent, 1 and 4, etc.

In the three-variable map we had cancellation when both the 0 and 1 values were shaded, since $0 + 1 = 1$. In the four-variable map both C and D are represented horizontally, hence four horizontal squares are required to cancel both the C and D values. If these four squares were opposite the AB 11 section, it would designate AB only. If the two lower hori-

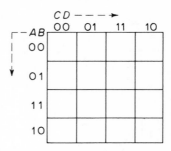

Fig. 7-16. A basic four-variable map.

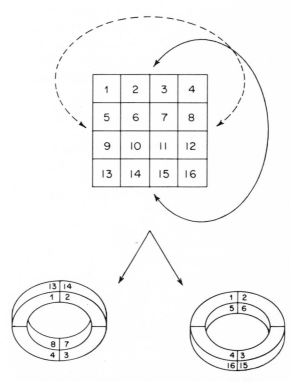

Fig. 7-17. Adjacency factors in four-variable map.

zontal sections are notated, as in Fig. 7-18, the *CD* combination is cancelled, leaving $AB + A\bar{B}$ which equals *A* only.

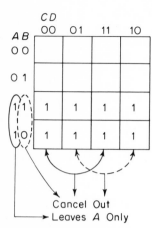

Similar logic applies to 1's in a vertical array as shown in Fig. 7-19. Here the *AB* combination cancels out, as well as the *D* value, leaving only *C*. Maps for *B* and *D* are shown in the upper portion of Fig. 7-20. For negated variables, the 1's formation is opposite to that for the complements, as can be seen by comparing the *A* value in Fig. 7-18 with the $\bar{A}$ at the lower left of Fig. 7-20. Note, also, how the $\bar{B}$ occupies horizontal rows not used for *B*.

Fig. 7-18. Formation of *A* in four-variable map.

Maps for $\bar{C}$ and $\bar{D}$ are shown at the top of Fig. 7-21. Since adjacent 1's are required for cancellation, eight 1's are involved as for the other single literals. Thus, if a single 1 appears in a square the complete four-variable term is indicated, as shown in the lower two maps. In the lower left map the 1 is opposite the 11 for

AB and below the 01 for *CD*, forming 1101, or $AB\bar{C}D$. Similarly, the single 1 in the lower right map is opposite the 00 for *AB* and below the 10 for *CD*, yielding

$$0010 = \bar{A} \cdot \bar{B} \cdot C \cdot \bar{D}.$$

Adjacency factors are illustrated in Fig. 7-22. In the upper left map the top and bottom 1's are considered adjacent, because of the wrap-

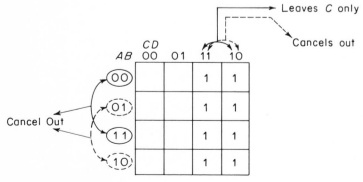

Fig. 7-19. Formation of *C*.

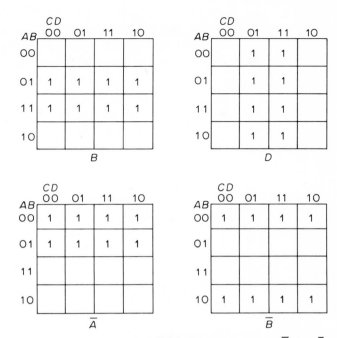

Fig. 7-20. Formation of B, D, Ā, and B̄.

around factor. The A variable is cancelled $(\bar{A} \cdot \bar{B}) + (A \cdot \bar{B})$ leaving only the $\bar{B}$. Since the 1's are below the 01 CD column we have $\bar{C}D$, which, combined with $\bar{B}$ provides us with $\bar{B} \cdot \bar{C} \cdot D$. The same factors apply to the upper right map, where adjacency occurs at the sides, cancelling the C and leaving $\bar{A}B\bar{D}$.

Examples of immediate adjacency are shown in the lower drawings. In the lower left map the 1's are below 01 and 11, cancelling the C variable. Since the 1's are opposite the 01 for the AB column, we obtain $\bar{A}BD$. For the map at the lower right double cancellation occurs, one for the vertical C and the other for the horizontal A to leave BD. Compare this with the B and D shown in Fig. 7-20, noting that when the two variables are entered on a single map, all squares not coincidental are left blank.

How a four-variable expression is entered into a map has already been shown in Fig. 7-21. As with the three-variable maps, the individual variables are indicated by a 1 placed into the appropriate square which satisfies the binary logic involved. The method for entering the expres-

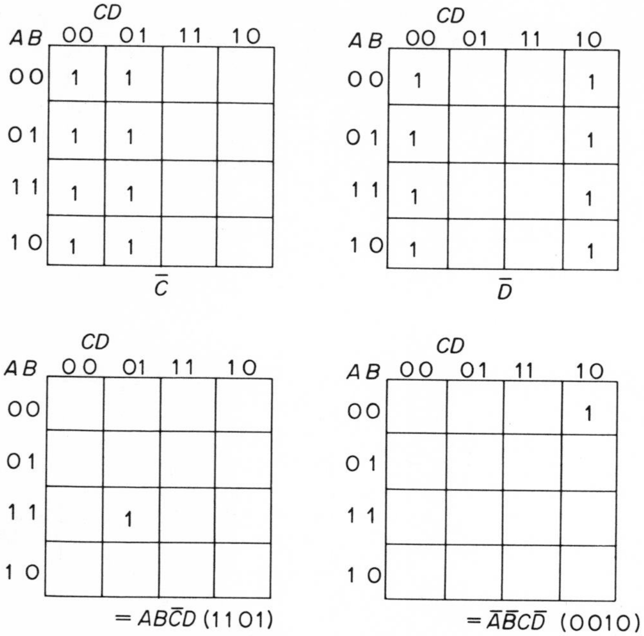

Fig. 7-21. Formation of $\overline{C}$, $\overline{D}$, $AB\overline{C}D$, and $\overline{A}\overline{B}C\overline{D}$.

sion $(\overline{A}\cdot B\cdot\overline{C}\cdot\overline{D})+(\overline{A}\cdot B\cdot\overline{C}\cdot D)+(A\cdot B\cdot\overline{C}\cdot\overline{D})+(A\cdot B\cdot C\cdot D)$ is shown in Fig. 7-23. The resultant adjacency of the four 1's indicates the simplified resultant as $B\overline{C}$, since the A and D are cancelled by the $0+1$ formations.

If we enlarge the foregoing expression to include $\overline{A}BCD+\overline{A}BC\overline{D}$, the two new terms, when added to our previous map, form a new map as shown in Fig. 7-24. Since we now have a complete horizontal row of 1's, the CD values for this row cancel, leaving $\overline{A}B$. Our original block of four is still present and gives $B\overline{C}$. Thus, we now have $B\overline{C}+\overline{A}B$ which, when simplified, gives $B(\overline{A}+\overline{C})$.

From the foregoing discussions and illustrations we can formulate the following rules:

Any block of four 1's cancels one literal in AB and one in CD. (7-1)

Any horizontal row of four 1's cancels CD. (7-2)

Any vertical row of four 1's cancels AB. (7-3)

140

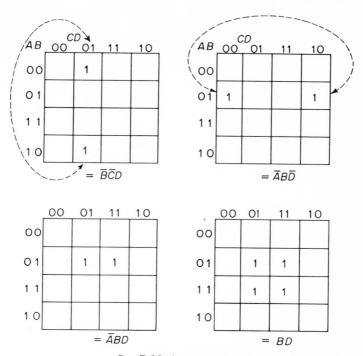

Fig. 7-22. Interpretation of mapped expression.

When single 1's are present and not linked by the adjacency factor (see Fig. 7-21) they form the full four-variable expression. Thus, if such a single 1 is present in a map containing other separate 1's, it must be represented in the final expression derived from the map. This is shown in the second drawing in Fig. 7-24. Here, the four 1's forming the block still represent $B\overline{C}$. The corner 1, however, indicates $\overline{A} \cdot \overline{B} \cdot C \cdot \overline{D}$ and, since it is not linked to any other 1 by the adjacency factor, the expression it represents must be included: $(B \cdot \overline{C}) + (\overline{A} \cdot \overline{B} \cdot C \cdot \overline{D})$.

Thus, each 1 in a map must be used at least once. If it is adjacent to other 1's it will, of course, be used in conjunction with the adjacent 1's again.

When a single square separates two 1's as shown in the upper left map in Fig. 7-25, the CD cancels, leaving only $\overline{A}B$. This occurs because the vertical 00 and 11 columns form $(\overline{A} \cdot B \cdot \overline{C} \cdot \overline{D}) + (\overline{A} \cdot B \cdot C \cdot D)$ wherein the $\overline{C} + C$ and $\overline{D} + D$ each are equivalent to 1 $(\overline{A} \cdot B \cdot 1)$, etc. For the single-square separation shown in the upper right map, we again cancel the CD, leaving AB as shown. Similarly, for vertical 1's sepa-

141

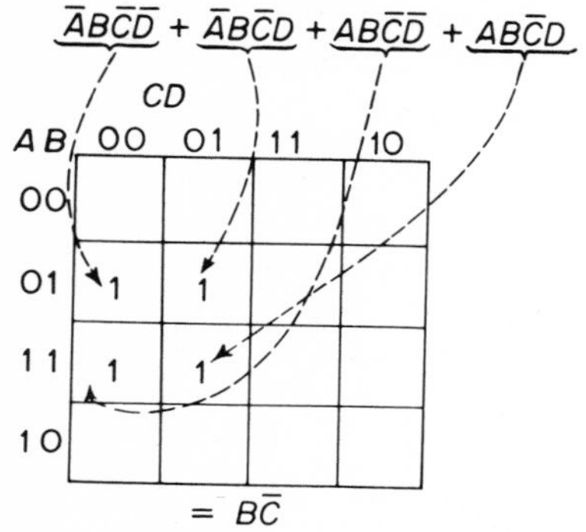

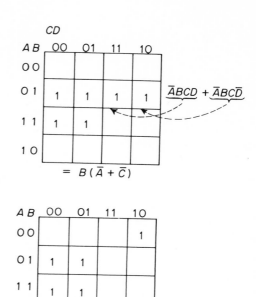

$\overline{ABCD} + \overline{A}BC\overline{D}$

$= B(\overline{A} + \overline{C})$

$\overline{AB}C\overline{D} + \overline{A}B\overline{C}D + AB\overline{C}D + AB\overline{C}D$

$= B\overline{C}$

$= B\overline{C} + \overline{A}\,\overline{B}C\overline{D}$

Fig. 7-23. Expression entry on four-variable map.

Fig. 7-24. Enlarged versions of Fig. 7-23.

rated by a single square as shown in the map at the lower left, the AB cancels, leaving $\overline{C}D$. Here, the AB column forms

$$(\overline{A}\cdot\overline{B}\cdot\overline{C}\cdot D) + (A\cdot B\cdot\overline{C}\cdot D)$$

and the AB values cancel. In the lower right map cancellation again occurs for AB, leaving CD. Thus, the following rule applies:

A separation of two 1's by a single square cancels the variables represented by the right-angle columns. (7-4)

Several examples of this rule are shown in Fig. 7-26. At the upper left the two 1's beneath the CD (11) cancel the AB, leaving CD. Horizontally, however, two adjacent 1's are opposite the AB column, indicating the term ABC. Thus, $CD + ABC = C(AB + D)$. In the upper right map, the horizontal 1's indicate $\overline{A}\cdot\overline{B}$ and the vertical 1's show $\overline{C}\cdot D$ for the complete expression $(\overline{A}\cdot\overline{B}) + (\overline{C}\cdot D)$, which is in its minimum form.

142

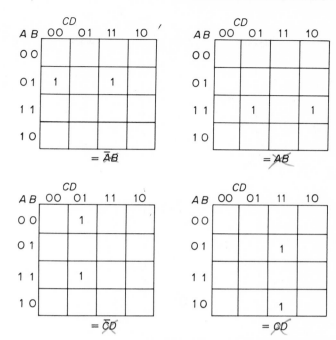

Fig. 7-25. Separation by one square.

Dual adjacencies exist in the two lower maps in Fig. 7-26. The lower right map is similar to the lower left because of the wrap-around factor which, in effect, makes the lower 1 adjacent to the upper 1. Both the vertical and horizontal adjacencies must be considered when deriving the simplified expression. For the lower left map the horizontal 1's yield ABC and the vertical 1's indicate ACD. Combining we have $ABC + ACD = AC(B + D)$. In the lower right map the horizontal 1's give $\overline{A} \cdot \overline{B} \cdot \overline{C}$ and the vertical (adjacent wrap-around) yield $\overline{B} \cdot \overline{C} \cdot \overline{D}$. Combining, we have $(\overline{A} \cdot \overline{B} \cdot \overline{C}) + (\overline{B} \cdot \overline{C} \cdot \overline{D}) = \overline{B} \cdot \overline{C}(\overline{A} + \overline{D})$.

The rule for such adjacencies is:

Derive the expression for the adjacent horizontal 1's; then use the 1 that links to the vertical 1 for deriving the expression for the vertical adjacencies.　　　(7-5)

Additional examples are shown in Fig. 7-27. Here three maps have three 1's in a row. The rule here is similar to rule (7-5):

143

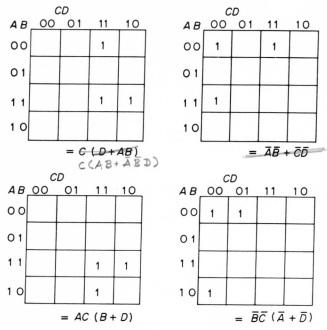

= C (D + AB)
    C (AB + ĀB̄D)

= ĀB̄ + C̄D̄

= AC (B + D)

= B̄C̄ (Ā + D̄)

Fig. 7-26. Representative examples (Group 1).

Use the center 1 to form a pair with an adjacent one and
derive the expression. Repeat with the center one forming
a pair with the remaining 1.

(7-6)

In the upper left map the original expression was:

$$(\overline{A}\cdot B\cdot \overline{C}\cdot D) + (A\cdot B\cdot \overline{C}\cdot D) + (A\cdot \overline{B}\cdot \overline{C}\cdot D).$$

Using the center 1 with the upper 1 we obtain $B\overline{C}D$. When the center 1
is used with the lower one to form a pair we obtain $A\overline{C}D$. Thus, the
combined terms form $B\overline{C}D + A\overline{C}D = \overline{C}D(A + B)$.

When the expression $(\overline{A}\cdot \overline{B}\cdot \overline{C}\cdot D) + (\overline{A}\cdot \overline{B}\cdot C\cdot D) + (\overline{A}\cdot \overline{B}\cdot C\cdot \overline{D})$ is
mapped we obtain a row of three 1's again as shown in the upper right.
Using the center 1 with the left 1 we obtain $\overline{A}\cdot \overline{B}\cdot D$. The center 1 in
conjunction with the right 1 produces $\overline{A}\cdot \overline{B}\cdot C$. Combining these and
simplifying we have $(\overline{A}\cdot \overline{B}\cdot D) + (\overline{A}\cdot \overline{B}\cdot C) = \overline{A}\cdot \overline{B}(C + D)$.

The map at the lower left is for the equation

$$(\overline{A}\cdot B\cdot \overline{C}\cdot \overline{D}) + (\overline{A}\cdot B\cdot \overline{C}\cdot D) + (\overline{A}\cdot B\cdot C\cdot \overline{D}).$$

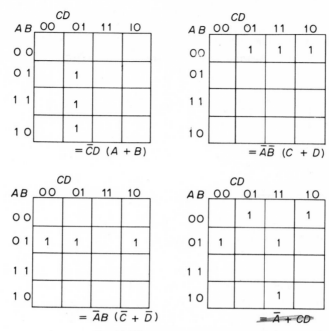

Fig. 7-27. Representative examples (Group 2).

Note that rule (7-6) applies here because of the adjacency factor relating to the right edge and left edge 1's. Even though there is a single square between 1's, rule (7-4) does not apply because *three* ones are present in a single row instead of *two*. The left pair of 1's forms the variable combination $\bar{A}B\bar{C}$. When the 1's at each edge are paired for adjacency we obtain $\bar{A}B\bar{D}$. Combining and simplifying produces

$$\bar{A}B\bar{C} + \bar{A}B\bar{D} = \bar{A}B(\bar{C} + \bar{D}).$$

The lower right map was formed from the equation

$$(\bar{A} \cdot \bar{B} \cdot \bar{C} \cdot D) + (\bar{A} \cdot \bar{B} \cdot C \cdot \bar{D}) + (\bar{A} \cdot B \cdot \bar{C} \cdot \bar{D})$$
$$+ (\bar{A} \cdot B \cdot C \cdot D) + (A \cdot \bar{B} \cdot C \cdot D).$$

Here only rule (7-4) applies. For the upper pair of horizontal 1's we obtain $\bar{A} \cdot \bar{B}$ only since the CD values are cancelled. For the second pair of horizontal 1's we obtain $\bar{A}B$. The second 1 in this row is also paired with the lower 1. This vertical resultant yields CD only because of the cancellation of AB. Thus, the combined expression is

145

$$(\overline{A} \cdot \overline{B}) + (\overline{A} \cdot B) + (C \cdot D).$$

Applying the distributive law we obtain the simplified form $\overline{A} + CD$. Two additional examples are shown in Fig. 7-28, with adjacencies

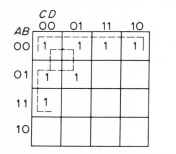
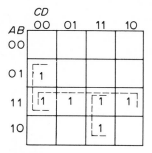

Fig. 7-28. Representative examples (Group 3).

indicated. In the left map the four horizontal 1's cancel the CD, leaving $\overline{A} \cdot \overline{B}$, (rule 7-2). The four 1's forming a block at the upper left of the map forms $\overline{A} \cdot \overline{C}$, (rule 7-1). The block also forms an adjacency with the lower one (rule 7-5) to yield $B \cdot \overline{C} \cdot \overline{D}$. Thus, the combination forms $(\overline{A} \cdot \overline{C}) + (\overline{A} \cdot \overline{B}) + (B \cdot \overline{C} \cdot \overline{D})$ which simplifies to

$$\overline{A}(\overline{B} + \overline{C}) + (B \cdot \overline{C} \cdot \overline{D}).$$

In the second map the first vertical adjacency produces $B \cdot \overline{C} \cdot \overline{D}$. The four horizontal 1's yield AB, and the second adjacency gives ACD. The combination is $(A \cdot B) + (B \cdot \overline{C} \cdot \overline{D}) + (A \cdot C \cdot D)$.

$(A \cdot C \cdot D)$

FIVE- AND SIX-VARIABLE MAPS

A number of variations will be found in the methods for depicting Karnaugh maps. The four-variable maps, for instance, can be shown with the AB values along the horizontal columns and the CD in the vertical. Similarly, squares can be shaded or 1's used. In some instances x's are also employed. There is little difference in the simplification procedures, however, and the choice of map configuration is unimportant. Obviously the maps used by a particular individual are those to which he has become accustomed.

A number of methods will also be found in the formation of five-

and six-variable maps. Those based on the three- and four-variable maps with which one is familiar are the easiest to learn. A five-variable map can be constructed by adding an additional 16 squares to the four-variable map as shown in Fig. 7-29. Note that the Gray-code numbering has been expanded along the horizontal columns to accommodate the additional literal.

Notation is similar to that discussed for the four-variable maps. If we wish to indicate $A \cdot B \cdot \bar{C} \cdot \bar{D} \cdot E$, we will place a 1 opposite the 11 of the

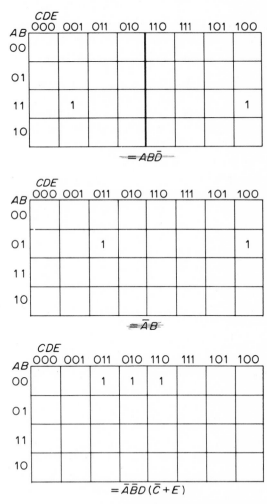

Fig. 7-29. Five-variable maps.

147

AB designation and below the 001 of the *CDE*. Similarly, $A \cdot B \cdot C \cdot \bar{D} \cdot \bar{E}$ is depicted by again placing a 1 opposite the 11 designation of the *AB* column, but now below the 100 vertical column designation of *CDE*. As shown in Fig. 7-29, these two terms produce the simplified expression *ABD̄*. This value is the resultant, because along the horizontal 11 row we find that *C* and *E* cancel because of the $1 + 0$ factor, leaving 11 for the *AB* designation, and two 0's for the *D̄* value.

In the second map in Fig. 7-29, the *CDE* value cancels because of the opposing 011 and 100 numbers, leaving only *ĀB*. For the third map, rule (7-6) as used for the four-variable maps applies here also. The first two 1's yield $\bar{A} \cdot \bar{B} \cdot \bar{C} \cdot D$. When the center 1 is linked to the other we obtain $\bar{A} \cdot \bar{B} \cdot D \cdot \bar{E}$. Thus, we have

$$(\bar{A} \cdot \bar{B} \cdot \bar{C} \cdot D) + (\bar{A} \cdot \bar{B} \cdot D \cdot \bar{E}) = \bar{A} \cdot \bar{B} \cdot D(\bar{C} + E).$$

The rules of adjacency which applied to the four-variable map also apply to each 16-square section of the five-variable map. In addition, any two 1's in a single horizontal column are considered adjacent if they are equidistant from the center line separating the two divisions. This can be visualized by imagining the map as folded in half along the vertical center line between the two sections. When folded, one

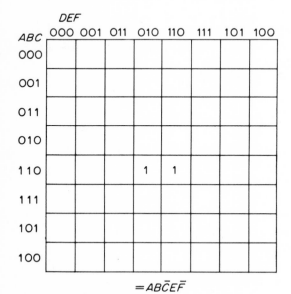

$$= AB\bar{C}E\bar{F}$$

Fig. 7-30. Six-variable map.

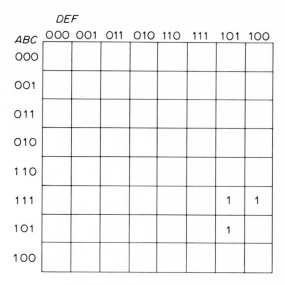

Fig. 7-31. Adjacencies in six-variable map.

section is in back of the other and coinciding back-to-back squares are considered adjacent.

A six-variable map can be constructed with 64 squares as shown in Fig. 7-30. Here the extended three-bit Gray-code notation is used for both the vertical and horizontal columns as shown. Assume we wished to map the expression $(A \cdot B \cdot \bar{C} \cdot \bar{D} \cdot E \cdot \bar{F}) + (A \cdot B \cdot \bar{C} \cdot D \cdot E \cdot \bar{F})$. The first member would be placed opposite the 110 of ABC and below 010 of DEF so that 110010 represents $A \cdot B \cdot \bar{C} \cdot \bar{D} \cdot E \cdot \bar{F}$. The second member requires a 1 adjacent to the first, as shown, to indicate 110110 for $A \cdot B \cdot \bar{C} \cdot D \cdot E \cdot \bar{F}$. Thus, the adjacent ones cancel the D literal leaving $AB\bar{C}EF$ as the final form.

The map shown in Fig. 7-31 is for the expression

$$(A \cdot B \cdot C \cdot D \cdot \bar{E} \cdot F) + (A \cdot B \cdot C \cdot D \cdot \bar{E} \cdot \bar{F}) + (A \cdot \bar{B} \cdot C \cdot D \cdot \bar{E} \cdot F).$$

The horizontal adjacencies yield $ABCD\bar{E}$ and the vertical adjacencies produce $ACD\bar{E}F$. This reduces to $ACD\bar{E}(B + F)$.

Maps could also be constructed for seven variables, though the number of squares involved makes for an unduly complex system. Even with the six-variable map all the adjacency factors must be carefully considered in multiterm expressions.

Questions and Problems

1. What is the advantage of the map method over Boolean-algebra manipulations for obtaining simplification?

2. Why is the Gray-code numbering system used in Karnaugh maps?

3. Prepare a three-variable map and obtain the simplified resultant for the expression:

$$(\bar{A} \cdot \bar{B} \cdot C) + (\bar{A} \cdot B \cdot C) + (\bar{A} \cdot B \cdot \bar{C}) + (A \cdot B \cdot C) + (A \cdot \bar{B} \cdot C).$$

4. Prepare a map and indicate the simplified expression for:

$$(A \cdot \bar{B} \cdot \bar{C}) + (\bar{A} \cdot B \cdot \bar{C}) + (A \cdot B \cdot \bar{C}) + (A \cdot \bar{B} \cdot C) + (A \cdot B \cdot C).$$

5. From the map in Fig. 7-32, derive the original expression as well as the simplified version.

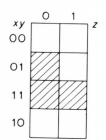

Fig. 7-32. Illustration for Problem 5.

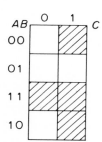

Fig. 7-33. Illustration for Problem 6.

6. From the map in Fig. 7-33, derive the original expression as well as the simplified version.

7. Draw a map and obtain the simplified expression from the following truth table:

A	B	C	Output
0	0	0	1
0	0	1	1
1	1	0	1
1	0	1	0
0	1	1	0
0	1	0	1
1	0	0	1
1	1	1	0

8. Draw a map and obtain the simplified expression from the following truth table:

A	B	C	Output
1	0	0	1
0	0	1	1
1	0	1	1
1	1	0	0
0	1	1	0
0	1	0	0
1	1	1	0
0	0	0	0

9. Prepare minterm and maxterm maps from the C_o output of the adder truth table shown in Chap. 5. Show how each table yields the same simplified term.

10. Prepare a four-variable map and obtain the simplified resultant from the expression: $ABCD + ABC\bar{D} + A\bar{B}CD + A\bar{B}C\bar{D}$.

11. Prepare a map and obtain from it the simplified term for the expression:

$$(\bar{A} \cdot \bar{B} \cdot C \cdot D) + (\bar{A} \cdot \bar{B} \cdot C \cdot \bar{D}) + (\bar{A} \cdot B \cdot \bar{C} \cdot \bar{D}) \\ + (\bar{A} \cdot B \cdot \bar{C} \cdot D) + (\bar{A} \cdot B \cdot C \cdot D) + (\bar{A} \cdot B \cdot C \cdot \bar{D}).$$

12. What was the original expression for Fig. 7-34, and what is the simplified resultant?

13. Draw a map and derive the simplified resultant for the expression $(\bar{A} \cdot B \cdot C \cdot D) + (A \cdot \bar{B} \cdot \bar{C} \cdot D) + (A \cdot \bar{B} \cdot C \cdot D)$.

14. What was the original expression for Fig. 7-35, and what is the simplified resultant?

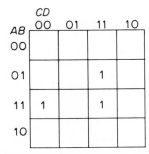

Fig. 7-34. Illustration for Problem 12.   Fig. 7-35. Illustration for Problem 14.

15. Prepare a map from the following truth table and obtain the simplified resultant:

A	B	C	D	Output
0	0	1	0	1
0	1	1	1	1
0	1	1	0	1

All other combinations = 0.

16. Prepare a map and obtain the simplified resultant for the expression: $(\bar{A} \cdot \bar{B} \cdot C \cdot D) + (A \cdot \bar{B} \cdot C \cdot D) + (A \cdot \bar{B} \cdot C \cdot \bar{D})$.

17. Map the following expression and obtain the minimum form: $AB\bar{C}D + ABCD + ABC\bar{D}$.

18. Map the following and obtain the minimum form:

$$(A \cdot B \cdot \bar{C} \cdot \bar{D}) + (A \cdot B \cdot C \cdot D) + (A \cdot B \cdot C \cdot \bar{D}).$$

19. From the map in Fig. 7-36, derive the original expression as well as the minimum form.

20. Draw a map representative of BC; then expand it to show $\bar{A}B + BC$.

21. From the map in Fig. 7-37, obtain the simplification.

22. Draw a five-variable map and obtain the simplified resultant for the expression: $(\bar{A} \cdot \bar{B} \cdot \bar{C} \cdot D \cdot E) + (A \cdot \bar{B} \cdot \bar{C} \cdot D \cdot E) + (A \cdot \bar{B} \cdot \bar{C} \cdot D \cdot \bar{E})$.

23. How many squares are there in a six-variable map? - 64

AB \ CD	00	01	11	10
00				1
01	1	1	1	1
11	1			
10				

Fig. 7-36. Illustration for Problem 19.

AB \ CD	00	01	11	10
00			1	
01				
11	1		1	
10				

Fig. 7-37. Illustration for Problem 21.

8

FLIP-FLOPS, PARALLEL ADDERS, AND STORAGE

INTRODUCTION

Because the sequential switching factors covered in Chap. 9 and Chap. 10 involve the storage characteristics of the flip-flop circuit, this topic is introduced in this chapter.

Also, in Chap. 5, it was pointed out that switching systems are used extensively in computer design, as well as in other industrial applications. Serial adders formed by switching circuitry were covered and the basic numbering system discussed. In this chapter the parallel adders are included, as well as associated switching systems, binary-coded decimal forms, shift registers, and computer storage.

THE FLIP-FLOP

The flip-flop circuit is widely used in computers and other systems as a bistable switching device. It also serves as a temporary storage for a binary representation for as long as power is supplied to the circuit. The flip-flop, as well as the logic AND, OR, and NOT circuits discussed previously are interconnected to perform the switching functions required, and are used over and over again in any particular computer or control system.

153

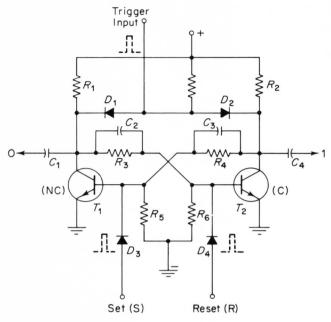

Fig. 8-1. Flip-flop circuit.

A typical flip-flop circuit is shown in Fig. 8-1. As with the other logical circuits discussed earlier, variations in design will be encountered for the ones used in industry. Since, however, the basic logic functions are identical for all, the one shown will suffice for an analytical description of its characteristics.

The bistable flip-flop has two stable states; that representative of logic 1 and the other of logic 0. When a triggering pulse is entered, the state changes either from 0 to 1 or from 1 to 0. In Fig. 8-1, the 0 state is represented, with transistor T_1 in its nonconducting state and transistor T_2 conducting. The transistors shown are NPN types, though PNP could also be used. Similarly, positive pulses are shown to indicate logic 1, although negative logic could also be employed if such a design is preferable.

When transistor T_1 is nonconducting, no current flows through it nor through resistor R_1. Hence, the full positive voltage of the supply appears at the collector. For transistor T_2 the conducting state causes a voltage drop across resistor R_2 and hence a decrease in the voltage at the collector.

154

The positive voltage at the collector of T_1 is coupled through R_3 to the base of T_2, thus providing the necessary forward bias to maintain conduction (positive base and negative emitter). The decreased voltage at the collector of T_2 is close to the ground level (negative) and this (coupled through R_4) keeps T_1 in a nonconducting state.

When a positive pulse is applied to the trigger input it appears at each collector. Since the voltage at the collector of T_1 is already positive, no effect is felt. For T_2, however, the positive pulse at the collector raises collector voltage. This rise in potential is also felt through R_4 at the base of T_1, causing this transistor to conduct. When T_1 conducts, there is a voltage drop across R_1, and the T_1 collector voltage drops toward the negative ground level. This (through R_3) cuts off T_2. With T_2 nonconducting, and T_1 conducting, the second state has been attained, and the flip-flop now registers logic 1.

It is of importance to note the voltage changes which occurred at each collector, since any voltage *change* is felt through coupling capacitors C_1 and C_4. When T_1 changed from nonconducting to conducting, the collector voltage dropped, producing the equivalent of an output negative voltage change. Thus, from the collector output designated as "0" a negative-pulse output is obtained at the time the flip-flop state is changed from 0 to 1. At T_2, however, the collector voltage rose, producing an output pulse of positive polarity at the time the state is changed.

If another triggering pulse is now applied, the positive potential will again be applied to both collectors and will raise the low voltage at the collector of T_1. This increase is also felt at the base of T_2 and provides the necessary forward bias to bring T_2 back into conduction again. In turn, T_1 is cut off, and the flip-flop now is in the 0 state again. During this change from 1 to 0, however, a positive voltage rise occurred at the collector of T_1 and produced a positive-pulse output from C_1. At C_4, however, the drop in collector voltage produces a negative-polarity output pulse.

From the foregoing it is evident that successive triggering pulses will change the state of the flip-flop from 0 to 1, from 1 to 0, and from 0 to 1 again, etc. Note, however, that two other pulse inputs are provided: the set input (S) and the reset (R). The set (S) input is used only for special purposes to place the flip-flop in the 1 state when it is at the 0 state. Successive pulse inputs at (S) will not trigger the stage once it is in the 1 state. A positive pulse at (S) is effective only when the base

155

of T_1 is negative (transistor nonconducting). Once the transistor conducts, its base is positive for an NPN type, hence the application of a positive pulse from the set line has no effect.

The reset input is used for clearing or resetting the flip-flop when it is in the 1 state. During this time T_2 is nonconducting, hence its base is negative (reverse bias). The positive-reset pulse applies the necessary forward bias to cause conduction and hence the flip-flop reverts to its logic-0 state. Once at 0, however, additional reset inputs have no effect, because the base is already positive. The reset input is useful for clearing a sequential train of flip-flops to erase a representative binary number. It will clear only those stages in the 1 state. Since it will not affect those stages in the 0 state, they will remain at 0 to correspond to the cleared stages.

Binary Counter

When flip-flop stages are connected together successively, they form counters. (Such counters are also referred to as *registers* in computers.) Block representations of flip-flops are shown in Fig. 8-2 where the necessary interconnections for forming counters are also illustrated. Note that the "0" output from one flip-flop is connected to the trigger input of the next stage. As shown at the top, if a single pulse is applied to the flip-flop at the right, it changes the state of this flip-flop to 1. During the change a negative pulse is produced at the "0" output and coupled to the next stage. The negative pulse is ineffectual, however, because the diodes in the triggering circuit will not pass this polarity. Thus, the flip-flop chain represents a binary one, 0001.

As long as no other triggering input pulse is applied, the flip-flop stages will retain this one (provided that dc power is maintained on the circuitry). Thus, such counters can also be considered as *temporary storage* (*memory*) devices. In computer usage, a single counter may have 12, 16, 24, or more flip-flop stages.

If another pulse is now applied, as shown in the counter representation below the top one, it will trigger the right stage to 0. At this time a positive pulse is obtained from the output and applied to the trigger input of the next stage. Now, the polarity is correct for passing through the diodes in the triggering circuit. (Such diodes are often called *steering diodes* because they, in effect, steer the input pulse to the proper

156

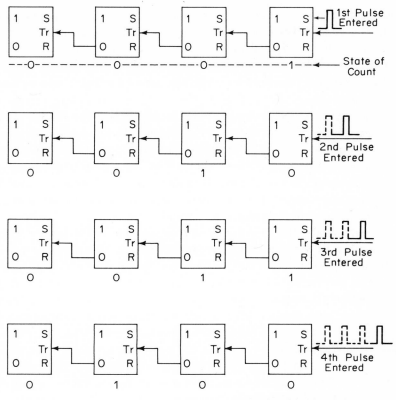

Fig. 8-2. Binary counter function.

flip-flop section for triggering.) When the positive pulse enters the second (from right) flip-flop, it places this stage in the 1 state. Now the flip-flop train registers 0010 to represent binary 2. (Flip-flop sequence is shown from right to left to indicate the successive representation, in binary form, of the input pulses.)

For a third pulse input, the first stage trips to logic 1 again. However, no positive output pulse is obtained, and there is no effect on the second stage, since only a positive pulse represents a 1 input (zero or negative voltage represents logic 0). Now the counter displays 0011 which equals binary 3. For the fourth pulse input, as shown at the bottom of Fig. 8-2, the count displays binary 0100 (base-ten 4). When the first stage triggers to 0, a positive-output pulse triggers the second stage to 0. This stage, in turn, produces a positive-output pulse and triggers the third stage to 1. When this stage changes from 0 to 1,

157

however, a negative-output pulse is produced which has no effect on the third stage. Thus, for successive pulse inputs the counter will represent the binary equivalent of the number of pulses entered. As such it accumulates counts and hence is also known as an *accumulator*. With thirty or forty flip-flop stages, the count can reach a magnitude in the millions.

The counter representation shown in Fig. 8-2 is also known as an *up counter* since it counts in ascending order. In some special applications, however, the counter can be designed to count down (*down counter*) by taking the output from the 1 side of the flip-flop and applying it to the trigger input of the next stage. If each flip-flop is placed in the 1 state, successive input pulses will progressively decrease the binary count representation of the counter.

Counters also have the ability to *scale*, since a positive output pulse (logic 1) is obtained for every two input logic 1's. Thus, if eight pulses are applied to the first stage (logic 1's), four will be obtained at the output, two logic 1's from the output of the second stage, and one from the output of the third stage.

DECADE COUNTER

Four flip-flop circuits, connected as a counter, can hold a maximum count of 15, or binary 1111. In many instances, however, the need is for a counter operating in *decade* fashion rather than in pure binary. In such operation four flip-flop stages count to 9 (1001) and trigger to 0 at the count of 10. Thus, such a group of four would feed another such combination and trigger the second group to 1 at the count of 10. This process permits operation in the base-ten mode for the counter while retaining the binary system for the individual stages of flip-flop circuits.

One method for converting a binary counter to a decade type is shown in Fig. 8-3. The flip-flop stages are connected in conventional up-count fashion, with the 0 outputs applied to the trigger inputs of successive stages. Stages marked *B* and *C*, however, have their 1 outputs applied to an OR circuit as shown. The output from the OR circuit feeds one input to an AND circuit. The other AND circuit input is obtained from the 1 output of flip-flop *D*. The output of the AND circuit is applied to the reset inputs of stages *B* and *C*. Thus, when

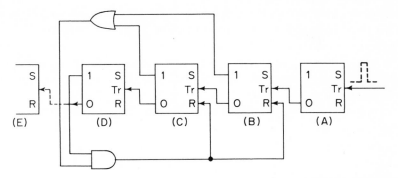

Fig. 8-3. Decade counter.

coinciding signals are applied to the AND circuit, stages *B* and *C* will be cleared if they are in the logic-1 state.

For this particular design, the OR and AND circuits are ineffectual up to and including the count of 9 (1001). When the tenth pulse is entered at the trigger input of the *A* stage, this flip-flop changes to 0 and triggers the next stage to 1. Thus, momentarily, the counter holds 1010. The 1 output from the *B* stage is applied to the OR circuit and, in turn, to the AND circuit. Since the *D* stage is also in the 1 state, its 1 output coincides with that of the *B* stage, thus an output is obtained from the AND circuit which applies a clearing pulse to the reset inputs of *C* and *D* stages. Stage *C* is already 0 and no change occurs here. Stage *B*, however, clears to 0 and in so doing trips *C* to the 1 state. Now, the 1 output from *C* and that from *D* again provide coincidence at the AND circuit. The latter applies clearing pulses again to *B* and *C*. Since *B* is now at zero, no change occurs. For *C*, however, the change is to the 0 state and an output pulse trips *D* to 0. Thus stages *A*, *B*, *C*, and *D* are cleared. As the latter clears, it trips stage *E* of the next decade counter to 1.

Successive decade counters are shown in Fig. 8-4. Here the flip-flop stages for a particular counter are shown stacked in a vertical plane. (It is assumed that each decade counter consists of the circuitry shown in Fig. 8-3.) With this arrangement the decade operational mode is in evidence and compares to the base-ten place system.

If we apply one pulse to the first counter (extreme right) the first-stage flip-flop will be in the 1 state and this decade counter thus represents base-ten 1. If three pulses are applied to the second decade counter, the *first two stages of this counter* each register 1, for a binary

159

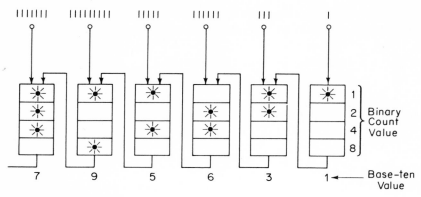

Fig. 8-4. Decade counters.

value of 3 (0011). With groups of pulses entered at the other decade counters as shown, the number representation for the entire system would be 795,631 as shown. If 9 more pulses were entered into the first decade counter, it would reach a 10 value and clear to 0. Now a pulse would be applied to the first stage of the second decade counter, changing its value to 4 in typical base-ten fashion.

Note that in the vertical plane the count representation is in pure binary. Horizontally, however, each representative value indicated by the individual decade counters forms the base-ten value. Each decade counter represents one place in the base-ten system and when the count for one place reaches 10, that place becomes 0, and 1 is added to the next place.

For a binary representation of the numbers stored in each decade counter it is necessary to code each decimal number in terms of its binary equivalent. Thus, this system is called *binary-coded decimal,* and for numbers above 9 differs from pure binary coding. In pure binary, for instance, 27 is represented as 11011. To show this number held by decade counters, however, the binary-coded decimal representation is 0010 0111 for 2 and 7.

Thus, we represent each base-ten *digit* by four binary bits (0's or 1's) to form an individual group. Thus, to represent 481, for instance, we use four bits for each binary-coded base-ten digit:

$$0100 \quad 1000 \quad 0001$$

When coding binary numbers in this fashion to represent base-ten

160

digits, the first 10 (0 to 9) would be the same as pure binary. At the 10 count the separation into groups begins as shown in the table below.

Base-ten Number	Binary-coded Decimal Form	
00		0000
01		0001
02		0010
03		0011
etc.		
09		1001
10	0001	0000
11	0001	0001
12	0001	0010
13	0001	0011
etc.		
20	0010	0000
21	0010	0001
22	0010	0010
23	0010	0011
24	0010	0100
etc.		
83	1000	0011
84	1000	0100
85	1000	0101
etc.		

PARALLEL OPERATION

Counters can handle numbers in either serial or parallel form, as required. The adders discussed in Chap. 5 operated in serial mode, as did a number of other circuits previously discussed. Parallel transfer of number representations can be performed as shown in Fig. 8-5. In the static state, the flip-flop number representation is available from the 1 output and the complement from the 0 output. If positive pulses represent 1's, a logic 1 would be obtained from the 0 output when the stage is triggered to 0 as previously discussed.

If the 1 outputs are connected to AND circuits as shown, the system is ready for transfer of the numbers in the upper counter to the one below. A 1 input is applied to the gate trigger input and thus provides coincidence for every AND circuit receiving the proper polarity from

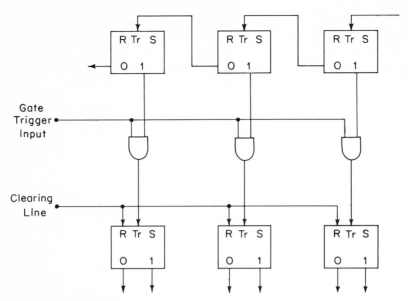

Fig. 8-5. Parallel number transfer.

its flip-flop stage. Thus, the AND gate opens and transfers the number to the trigger input of the lower counter stage. If an upper flip-flop is in its 0 state, the voltage at the 1 output is negative and no coincidence would occur at the AND circuit, and the flip-flop below it would, in essence, receive a 0 input. Thus, the lower flip-flop would not be triggered.

One form of parallel addition is shown in Fig. 8-6. Here, assume the counter already holds the binary value 001 which is to be added to 011 obtained from another counter or from storage. When the addition is to take place, a pulse is entered into the gate trigger and hence all input 1's are transferred to the individual stages of the counter in parallel form as was the case for the counters shown in Fig. 8-5.

When the rightmost 1 bit enters FF_1 at the right, it trips it to 0 and sends a triggering pulse into the delay line. (The reason for the delay line will be made evident later.) The second-place 1 input is applied to FF_2 and triggers it into the 1 state. When a flip-flop triggers from 0 to 1 there is no effect on the next stage, hence FF_3 remains at 0. Now, momentarily, the count is 010; the pulse from the delay line between FF_1 and FF_2 exits and triggers FF_2 from 1 to 0. This, in turn, triggers FF_3 into 1 for a final sum representation of 100.

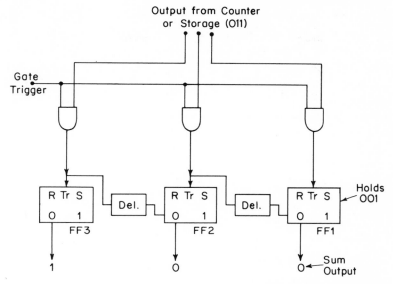

Fig. 8-6. Parallel adder.

In this particular system the delay lines are necessary to avoid double entry of digits and false addition. Without the delay lines, the triggering of FF_1 from 1 to 0 would have applied a triggering pulse to FF_2 at the same time that the second-place 1 entered from the top, resulting in a trigger to 1 for FF_2 without subsequent correction. Thus the false sum 010 would have resulted if no interstage delay were present.

SHIFT REGISTERS

Flip-flop type counters can be designed to shift a binary number a specific number of places for multiplication and division purposes, as well as for serial interchange of data between registers. As in conventional arithmetic, when binary numbers are multiplied, partial products are formed as shown below, and these are added together to form the final product.

$$
\begin{array}{r}
1011\ (11) \\
\times\ 101\ (05) \\
\hline
1011 \\
0000 \\
1011\ \\
\hline
110111\ (55)
\end{array}
$$

163

In the foregoing, each partial product is shifted left one place. Note the simplicity of binary multiplication: partial products are either zeros or composed of the multiplicand. Unlike conventional arithmetic, we need not remember multiplication tables, since $1 \cdot 1 = 1$ and $1 \cdot 0$ is 0.

For division, right shifting is involved, as shown by the following example:

$$
\begin{array}{r}
110 \\
101)\overline{11110} \\
\underline{101} \\
101 \\
\underline{101} \\
101 \\
\underline{101} \\
\overline{000}
\end{array}
$$

A method for modifying a counter to form a shift register is shown in Fig. 8-7. Here, interstage delay lines are used and the reset line

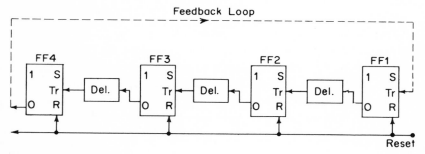

Fig. 8-7. Shift register.

employed to initiate the shifting process. Assume, for instance, that the rightmost flip-flop is in the logic-1 state. If a reset pulse is applied, this stage will clear and thereby produce an output pulse which would normally trigger the next stage to logic 1. The delay line, however, prevents the next stage from being triggered immediately. After the clearing pulse duration the signal in the delay line enters the next flip-flop and trips it to the 1 state. Thus, the initial number 0001 is now 0010 because of the shift to the left. Another reset pulse would shift the 1 to the next stage to produce 0100, etc.

The shifting process is not confined to a single bit, however. If the number 101 were present, the first reset pulse would clear all stages and in so doing a pulse from the right flip-flop and a pulse from the third-from-right would enter the delay lines. After the clearing pulse duration these signals would emerge from the delay lines and trip respective stages to cause the counter to register 1010.

A register would, of course, have more flip-flop stages than shown in Fig. 8-7. If the output from the last flip-flop is fed back to the first, a *ring counter* is formed, since the feedback loop forms an equivalent ring formation of the sequential stages. In such an instance any number held by the counter can be recirculated as desired.

The contents of one counter (register) can be shifted into another by feeding the output of the first into the input of the second, as shown in Fig. 8-8. The registers shown are eight-bit capacity types. Hence,

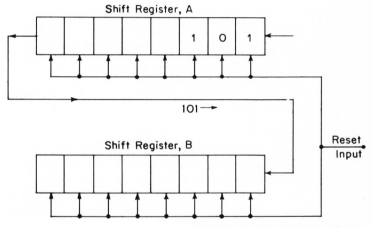

Fig. 8-8. Register data interchange.

eight reset pulses are required to shift a number from one register into the other.

If more than eight reset pulses are applied, the shifted number would undergo multiplication. Thus, if nine reset pulses are used, the 101 shown in register A would appear in register B as 00001010 which is double the original value of 5. If 10 reset pulses are used, the number becomes 00010100 which equals 20, or four times the original value. Similarly, if register A originally contained 1100 (12) the application of nine reset pulses would form 00011000 which equals 24, double the original value. If, however, only seven reset pulses were employed, division would occur. For a shift of only 7, the number appearing in register B is 00000110, (6), which represents division by 2. A shift of only 6 would produce 00000011 for a value of 3, a division of the original number by 4.

This multiplication and division by shifting does not replace standard arithmetical methods since the process is limited to the base-two system of 1, 2, 4, 8, 16, etc. Multiplication or division by 3, 5, 7, etc. can't be made in this fashion.

STORAGE SYSTEMS

There are various methods employed for storage of binary information. The registers discussed in this chapter are, in essence, storage devices since numbers will be held by such counters as long as power is applied to them. These devices are, however, temporary storage sections. Also required is a memory that will retain data after operating power is removed from the system.

Storage devices based on magnetic principles have been the most widely used in computer design, and consist of magnetic tapes, magnetic discs, toroids of ferrite (called *ferrite cores*), twisted wire, magnetic drums, etc. Magnetic cores, drums, and discs are rapid access (high-speed read-in and read-out) and are often referred to as *internal storage devices*, as opposed to the external types composed of slower-access magnetic tapes, punched paper tapes, and punched cards.

Because the computer only recognizes binary values of 0 and 1, a magnetized area can be assigned a 1 value and an unmagnetized area logic 0. In the drum and disc types, recording and reading heads are used to read in and read out binary data in similar fashion to the sound recording and playback methods used in home tape recorders. Ferrite cores are energized by the magnetic fields set up in wires strung through the core centers, as shown in Fig. 8-9.

Ferrite cores vary in size, with many having an outside diameter of less than 0.02 inch. They are wired into a frame often less than 2×2 inches square and containing over 4,000 individual cores. These frames, in turn, are stacked to conform to the number of bits capacity of the associated registers. Thus, if 16-bit registers are in use, the core-storage stacking may also be in the order of 16 for full magnitude number interchange between memory and counters.

A single-plane wiring is shown in Fig. 8-9. The vertical and horizontal wires identified by A to D and N to Q are for read-in purposes.

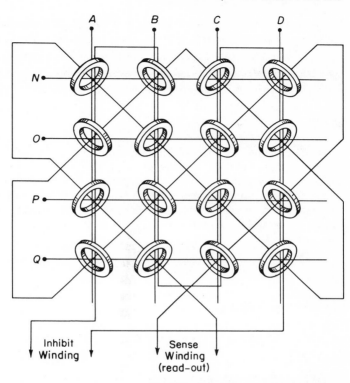

Fig. 8-9. Magnetic core storage system.

If a logic 1 is to be placed in the upper-right ferrite core, for instance, wires marked *N* and *D* would be energized. Each wire, however, only carries enough energy to provide one-half the magnetic field that would change the magnetic state of the core. Where the wires cross, the two fields add to provide full magnetic intensity to change the magnetic state of the core. The voltages applied are sometimes referred to as *half-set* voltages, since each only sets up the magnetic state to one-half necessary for tripping. Thus, even though the current flow through wire *N*, for instance, builds up magnetic fields at the center of the first three cores, such fields are not of sufficient amplitude to affect these cores. Similarly, the *D* vertical line also half-energizes all vertical cores at the right, but not sufficiently to change the magnetic state. Only where the wires intersect is the core state changed.

During read-out of the information, voltages of opposite polarity are applied to the lines for intersection at the desired core. The magnetic status of such a core is changed and the collapsing magnetic field induces a voltage on the read-out *sense* wire. Since the information is lost during read-out, the data are recirculated and reapplied to maintain permanent storage status. The *inhibit* winding assures proper core-state restoration without its having an effect on other cores by circulating a current which prevents (inhibits) other cores from changing their state of either 0 or 1.

NOTE. This chapter and Chap. 5 have summarized the general aspects of counters and storage. It is intended to supply sufficient background in these subjects so they can be related to the switching theory covered in other chapters. For a more comprehensive coverage and detailed explanations of these and other computer circuitry, reference should be made to the author's *Fundamentals of Electronic Computers* (*Digital and Analog*), Prentice-Hall, Inc., 1967.

Questions and Problems

1. Define the terms *set*, *reset*, and *trigger*, as they apply to a flip-flop circuit.

2. Why is a flip-flop considered only a temporary storage device?

3. Under what circumstances does the triggering of a flip-flop cause the triggering also of the next stage?

4. What are the purposes for *steering* diodes?

5. Explain briefly how a *decade* counter differs from a conventional binary *counter*.

6. Express the numbers 583 and 1068 in binary-coded decimal form.

7. Explain the purpose of the interstage delay lines in the parallel adder of Fig. 8-6.

8. Why are multiplication tables unnecessary when performing binary multiplication?

9. Explain the difference between a shift register and a ring counter.

10. Assume two 16-bit shift registers are in use, and a number is transferred from one to the other. By how many bits must the number

be shifted to retain its value? By how many bits to multiply the number by 4?

11. Name three types of magnetic storage devices and indicate the general method for read-in and read-out of information.

12. Explain what is meant by a *half-set* voltage in reference to ferrite-core storage.

13. What is the purpose for the *sense* wire in ferrite-core storage?

14. What is the advantage of ferrite-core storage over magnetic-tape storage?

15. Why are ferrite cores, magnetic drums, discs, and magnetic tape, considered permanent storage as opposed to the temporary storage devices such as flip-flops?

9

SEQUENTIAL CIRCUITS

Introduction

INTRODUCTION

A *sequential* switching system is formed when signal states are delayed or stored in memory devices prior to their influencing the final signal output. Hence, the output expressions are not representative of the Boolean logic functions that exist at the input terminals. In contrast, the combinational switching system produces an output directly determined by existing input combinations. For the sequential systems, the outputs are formed not only by the existing inputs, but also by switching states which have occurred earlier.

In *electrical* switching systems contact switches are used and combined to form logic AND and OR gates as described in Chap. 4, where they were compared with the electronic (diode, transistor) types. In such systems the use of relays for switching purposes forms sequential circuitry because of the switching delay or holding characteristics introduced by the solenoid (magnet coil). Similarly, the use of a delay line (a series of inductors and shunt capacitors) produces a sequential switching system.

In *electronic* systems the data-storage memory devices form sequential switching when used in conjunction with the combinational circuitry. Thus, a complete logic-switching system (either electrical or electronic) often utilizes the combinational and sequential circuitry simultaneously.

170

In this text the logic as it applies to the electronic sequential system is discussed, although the principles are also applicable to the electrical systems.

SYMBOLS AND TERMS

The inputs to sequential circuits are called *primaries* (unrelated to transformer terminology), and their memory characteristics are *secondaries*. Literals x, y, Y, and z are used to represent primaries and secondaries, including the negation sign ($\bar{x}$, $\bar{y}$, etc.) when necessary. The letter designations have the following representations:

$x_1, x_2, \ldots, x_n$ represent binary inputs of primaries

$z_1, z_2, \ldots, z_n$ represent binary outputs of primaries

$y_1, y_2, \ldots, y_n$ represent *present* states of secondaries (memory)

$Y_1, Y_2, \ldots, Y_n$ represent the next states of secondaries

The next state of the secondary (Y) is also called *excitation* or memory input. The y state of a secondary thus changes to Y after the excitation has been applied to the input of the storage device. These factors are shown in Fig. 9-1 for both the normal and negated forms.

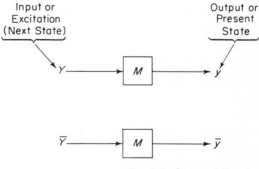

Fig. 9-1. Status of Y and y.

The complete sequential system is shown in Fig. 9-2. Additional x inputs and z outputs could, of course, be present if required. Similarly, there could be several Y excitation literals as well as y present-state

171

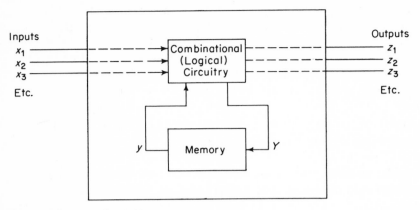

Fig. 9-2. Sequential switching system.

secondary representations. Where flip-flop circuits are used for the memory sections, the trigger input or the set and reset terminals are used for the Y excitation inputs, and the outputs represented by y and $\bar{y}$ as shown in Fig. 9-3. Thus, in certain applications only the trigger

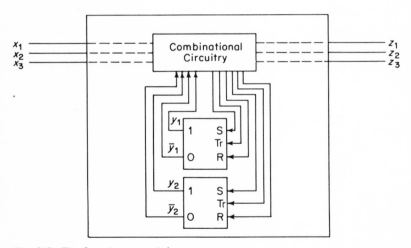

Fig. 9-3. Flip-flops in sequential system.

input would be used for the Y, while on other occasions the set (S) and reset (R) terminals would be employed (or combinations of all three as shown in Fig. 9-3).

As with combinational switching systems, the literals represent logical

172

0's or 1's. Instead of *A*, *B*, and *C*, etc. to designate separate inputs, subscripts are used with the *x* literal. For these, as well as the others, however, occasional use is made of $x = 0$, $z = 1$, etc., to give added clarity when needed. Sometimes expressions such as $z_1, z_2 = 01$ may be used where it is important that the specific values of the output literals be designated.

ILLUSTRATIVE TYPES

Sequential switching systems are found in a variety of forms, with each designed to meet a specific logical function. One representative example is the decade counter discussed in Chap. 8 (Fig. 8-3) as well as other systems using logical combinational switching with flip-flop circuitry.

Another example is shown in Fig. 9-4, where an AND circuit, a

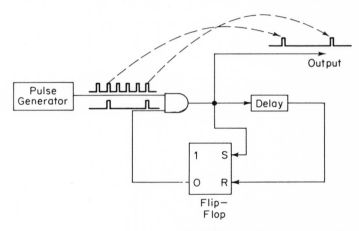

Fig. 9-4. Repetition-rate reduction.

delay line, and a flip-flop combine to form a pulse repetition-rate reduction system. A pulse generator produces a continuous train of pulses at a fixed pulse repetition rate. Initially, assume the flip-flop is in its logic 0 state and that at this time the 0-output line has a positive potential. This positive potential, applied to one input of the AND switch, provides coincidence with the first pulse applied to the AND gate from the pulse generator. Hence, a pulse output is obtained from

173

the AND circuit as shown. This pulse is also applied to the delay line as well as the set input of the flip-flop. Now the flip-flop is set to its 1 state and this changes the 0-output line toward the negative potential. Coincidence no longer occurs at the AND gate for several successive output pulses from the pulse generator.

When the pulse in the delay line finally emerges, it resets the flip-flop to 0 again, thus providing the positive-polarity output for coincidence again at the AND circuit. Consequently, another output pulse is obtained from the AND circuit. Again, this pulse also enters the delay line and at the same time trips the flip-flop to its *set* (logic-1) state. Eventually the delay-line pulse emerges and resets the flip-flop for a repetition of the entire process. For a specific delay, one pulse will be obtained from this circuit for every four from the pulse generator. For a shorter delay, there is less repetition-rate reduction from this sequential switching system.

An additional example of a practical sequential switching system is shown in Fig. 9-5. Assuming that positive logic is used, two diodes are

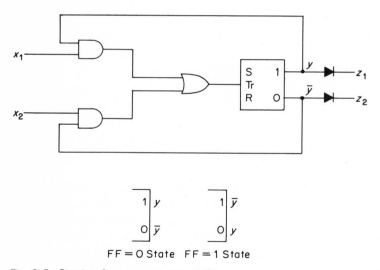

Fig. 9-5. Circuitry for $z_1 = x_2\bar{y}$; $z_2 = x_1\bar{y}$.

in series with the z_1 and z_2 lines to prevent a negative-pulse output from either during the switching cycle. The output logic is $z_1 = x_2\bar{y}$ and $z_2 = x_1\bar{y}$.

When the flip-flop is in its 0 state, the voltage level from the 1-output

174

line is toward the negative, hence no coincidence can occur for the x_1 input. The 0-output line, however, is at a positive-voltage level and this would provide coincidence for the x_2 input. Thus, an output pulse is obtained from the x_2 AND circuit and fed through the OR circuit to the flip-flop trigger input. The flip-flop now changes to its logical-1 state. Now the 0 output line is toward the negative potential and no longer provides coincidence for the x_2 input. The 1-output line, however, is positive and coincidence prevails at the x_1 input. The AND circuit output again feeds to the OR circuit and the flip-flop is triggered to its 0-state again, switching its input characteristics to the x_2 input once more.

In this example it was assumed that the 1-state representation of the flip-flop stage produced a voltage level from the 1-output line toward the negative, and the 0-output toward the positive. For the 0-*state* of the *flip-flop*, the 1-output line was positive, and the 0-output line negative, etc. These polarity changes must not be confused with the $y = 1$ and $\bar{y} = 0$ logic designations for the output lines. Had we used negative logic, where a negative voltage level represented logic 1 and a positive voltage indicated logic 0, it would not have changed the y designations for the flip-flop outputs. Negated literals do not imply negative voltage potentials, and negative logic should not be confused with logical negation. (See "Negative Logic and Negation," page 74 of Chap. 4.)

PULSE AND LEVEL SIGNALS

It must be remembered that logic gates can be opened or closed with steady-state voltage levels as well as pulse-type signals. This principle is illustrated in Fig. 9-6. At A the nonconducting (T_2) transistor of the flip-flop has the maximum positive potential at the collector and this is applied to one input of an AND switch. Thus, one AND input is in readiness to provide for an open-gate condition as soon as coinciding voltages are present at the other input. As shown in (A), if pulses representative of 101 are now applied to the other input of the AND circuit an output 101 is produced. A similar output would have been obtained had both inputs to the AND-gate terminals been pulses, such as (101)(101).

For the flip-flop shown in (B), transistor T_2 is conducting, hence there is a voltage drop across R_2 and a decrease in the collector voltage. The

175

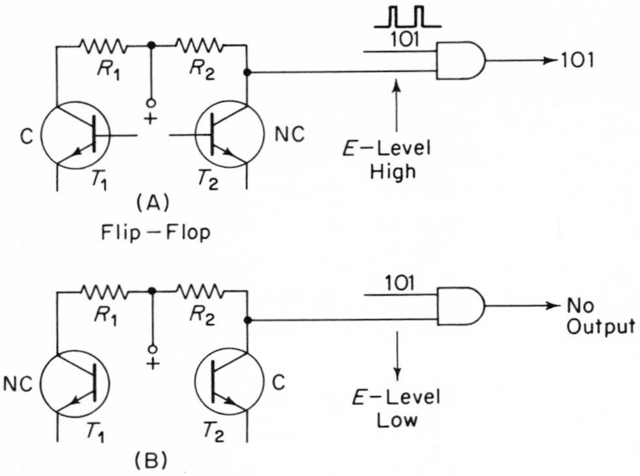

Fig. 9-6. Steady-state (level) switching.

reduced collector voltage is near the ground (negative) level and no longer provides the steady-state positive potential necessary for coincidence. The AND gate remains closed and no output is produced. Thus, the steady-state dc potentials present at the collectors represent *levels* of signal, rather than *pulses*. These factors must be taken into consideration in constructing diagrams and tables that describe the logical progressions applying to given sequential circuitry.

The diagrams and tables are necessary steps in the synthesis procedures for sequential circuits involving the initial problem statement, reduction, and final design. The steps required are given below, and the topics listed will be discussed individually in the remainder of this text.

1. The verbal presentation of the problem and the related specifications.
2. Flow-diagram representation and accompanying tables.
3. Reduction, if possible, of diagram-table representations.
4. Assignment of binary numbers representative of memory states of reduced diagrams.
5. Preparation of excitation maps and designing corresponding circuitry.

DIAGRAMS AND TABLES

Assume that a switching system is to have two states with two input signals. Only the second input signal is capable of changing a state.

176

If we assign S_1 and S_2 for state representation, and x_1 and x_2 for the signal inputs, the representative diagram would be as shown in Fig. 9-7A. Each state is represented by a large circle, and curved arrows indicate input functions and the transition in states taking place for inputs x_1 and x_2.

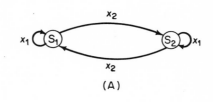

(A)

Note that for the S_1 circle a curved arrow emerges and loops back into the circle for x_1. This indicates that if an x_1 input is used at this state (S_1) there is no state change and the system remains in S_1. If, however, an x_2 input is used at S_1, the state changes to that indicated as S_2. If, at S_2, an input x_1 is applied, the state does not change. If, however, an x_2 input is used, the state changes to that of S_1 again as indicated by the curved arrow looping from S_2 to S_1. Because such a drawing represents transitions between states, it is sometimes called a *transition diagram*.

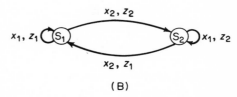

(B)

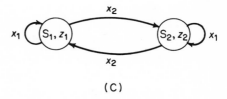

(C)

Fig. 9-7. Basic flow diagrams.

Signal logic flow is an equivalent designation, hence the term *flow diagram* has also been used.

If pulse-output signals are to be designated, they are positioned as shown in Fig. 9-7B. Here, input x_1 at state S_1 does not change the state, but an output z_1 is produced. If x_2 input is applied during the S_1 state, a z_2 output is obtained and the state changes to S_2. If x_1 is the input, the state remains the same and an output z_2 is again obtained. For an x_2 input at S_2 an output z_1 is obtained and the state changes to S_1. Thus, in (B), each z represents the *next output* obtained for a given input.

If *existing output* levels are to be designated, they can be placed next to the S_1 circles, or placed within them, as shown in Fig. 9-7C. Thus, z_1 within the S_1 circle indicates the level output present at S_1. Since an x_1

177

input does not change the S_1 state, there is also no output level change. For an x_2 input, however, the state changes to S_2 and its output level is z_2 as shown.

For the diagram shown in (B), the following table applies:

Present State	Next State x_1 x_2	Next Output x_1 x_2
S_1	S_1 S_2	z_1 z_2
S_2	S_2 S_1	z_2 z_1

Note that the horizontal row next to the *present-state* S_1 designation shows the changes which occur for the x_1 and x_2 inputs. For x_1 the state remains at S_1, while for an x_2 input the next state becomes S_2. For the present-state S_1 there is a z_1 output for an x_1 input, and a z_2 output for an x_2 input. Similar data are displayed for the horizontal row identified by the present-state S_2.

For the diagram in (C), only the two existing or present output levels are shown in the table:

Present State	Next State x_1 x_2	Present Output Level
S_1	S_1 S_2	z_1
S_2	S_2 S_1	z_2

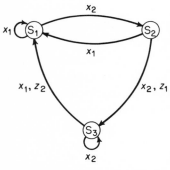

Fig. 9-8. Three-state flow diagram.

A three-state flow diagram with two inputs is shown in Fig. 9-8. An output z_1 is obtained only when an input x_2 is applied during state S_2, and output z_2 only occurs when state S_3 has an x_1 input applied. When in state S_1 an x_1 input does not change the state nor produce an output. An input of x_2 changes the system to S_2, again with no output. If an x_1 input is applied during S_2 the system reverts back to the S_1 state.

At S_2, an x_2 input not only changes the state to S_3 but also produces a z_1 output as shown. An x_2 input at S_3 has no affect on the state.

Since the z designations signify next outputs, and because only two

178

are involved, we can simplify the table by listing them with the next states, as shown:

Present State	Next State and Output	
	x_1	x_2
S_1	S_1	S_2
S_2	S_1	S_3, z_1
S_3	$S_1 z_2$	S_3

Several inputs can be associated with single arrow loops, as shown in Fig. 9-9. Here, for every state change with an x_1 input we obtain an

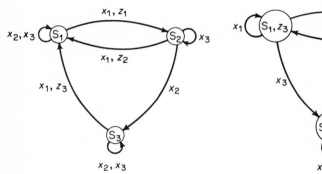

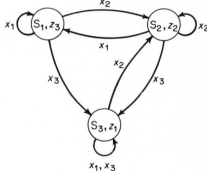

Fig. 9-9. Diagram with three inputs, outputs, and states.

Fig. 9-10. A three-input, three-level diagram.

output. There are three inputs, three outputs, and three levels. When in state S_1, an x_2 or an x_3 input does not change the state nor produce an output. At S_1, however, an x_1 input will produce a z_1 output and a change of state to S_2. At S_2 an x_1 input again produces an output, but it is z_2 and the state changes to S_1. For an output of z_3 the system must be in the S_3 state and an x_1 input applied, as shown. The accompanying table is shown below.

Present State	Next State and Output		
	x_1	x_2	x_3
S_1	S_2, z_1	S_1	S_1
S_2	S_1, z_2	S_3	S_2
S_3	S_1, z_3	S_3	S_3

179

Diagrams showing present levels can also have more than one input represented by a single arrow loop, as shown in Fig. 9-10. Note that at S_3 the level is z_1 and this state and level is maintained for an input of either x_1 or x_3. For an input of x_2, however, state S_2 is again reached, with a z_2 level. Also note that output level z_3 is present at S_1. The appropriate table is shown below.

Present State	Next State x_1 x_2 x_3			Present Output Level
S_1	S_1	S_2	S_3	z_3
S_2	S_1	S_2	S_3	z_2
S_3	S_3	S_2	S_3	z_1

Additional examples of flow diagramming and table construction follow.

REPRESENTATIVE EXAMPLES

Example 1. An output z is to occur only when a second x_1 input follows an x_2 input. The circuitry has three states, and for S_1 an input of x_1 initiates no state change. For S_2, an x_2 input initiates no state change. At S_3, the x_1 changes state to S_1.

The flow diagram for this problem is shown in Fig. 9-11. Note that only for the x_1 input at state S_3 is an output z obtained. Since this is the second x_1 input following an x_2 input, the diagram is valid. The problem statement did not restrict the x_2 input at S_3, and hence this input initiates a state change to S_2.

Starting at S_1 the input values are examined and their functions noted and set in table form. Successive analyses of inputs at S_2 and S_3 are set down to complete the table:

Present State	Next State and Output x_1	x_2
S_1	S_1	S_2
S_2	S_3	S_2
S_3	S_1, z	S_2

Example 2. An output z_1 is to occur at the second consecutive x_1 pulse input, and a z_2 output is to be obtained at the third consecutive x_1 input. At S_1 an x_2 input initiates no state change, and at S_3, an x_2 input causes no change of state.

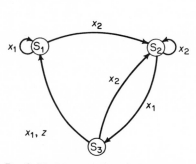

Fig. 9-11. Diagram for Representative Example 1.

Fig. 9-12. Diagram for Representative Example 2.

The flow diagram for this problem statement is shown in Fig. 9-12. While the problem doesn't state that an x_1 input at S_3 changes the state to S_1, such a transition is obvious in order to maintain the specification that each third x_1 input yields a z_2 output. The x_2 input at S_2 changes the state to S_1. The other x_2 inputs, as mentioned in the problem statement, initiate no state change, hence the arrow loop returns to the appropriate state circle as shown. The table preparation presents no difficulties.

Present State	Next State and Output	
	x_1	x_2
S_1	S_2	S_1
S_2	S_3, z_1	S_1
S_3	S_1, z_2	S_3

Example 3. When several output z's are encountered, it may be more expedient to list them under a separate table heading as was done in the table for Fig. 9-7B. Another example is shown in Fig. 9-13, where three inputs are involved plus four z_1 outputs and a single z_2 output. The following table applies to the diagram of Fig. 9-13:

181

Present	Next State			Next Output		
State	x_1	x_2	x_3	x_1	x_2	x_3
S_1	S_2	S_1	S_3	z_1		z_1
S_2	S_1	S_2	S_3	z_2		z_1
S_3	S_3	S_2	S_3		z_1	

Example 4. A circuit is to have four states and two inputs x_1 and x_2. At S_1 and S_2 the x_2 inputs do not change the state. At S_3 an x_2 input changes the state to S_1, and gives a z_1 output. At S_4, an x_1 input does not change the state. At S_1 three consecutive x_1 inputs produce z_2, z_1, and z_2 outputs. At S_4 an x_2 input gives a z_1 output and changes the state to S_1.

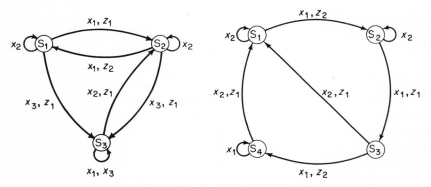

Fig. 9-13. Representative Example 3.   Fig. 9-14. Representative Example 4.

The four-state flow diagram for this problem statement is shown in Fig. 9-14. Again, since several outputs are present, it is more convenient to list them under a separate heading as shown in the table which follows:

Present	Next State		Next Output	
State	x_1	x_2	x_1	x_2
S_1	S_2	S_1	z_2	
S_2	S_3	S_2	z_1	
S_3	S_4	S_1	z_2	z_1
S_4	S_4	S_1		z_1

Example 5. A two-state system has three inputs. At S_1 and S_2 the x_1 input does not change state. At S_1 and x_2 input changes the state to S_2, where another x_2 input has no effect on changing state. At S_1 an x_3 input also changes the state to S_2. For an x_3 input at S_2, a z output is produced and the state changes to S_1 again.

Two diagrams are shown in Fig. 9-15 to illustrate the possible varia-

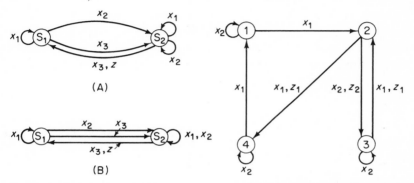

Fig. 9-15. Representative Example 5.   Fig. 9-16. Representative Example 6.

tions which may be encountered in flow-diagram construction. The x_1 and x_2 no-state-change inputs at S_2 can be indicated with separate loop arrows as shown in (A), or as a single loop arrow as shown in (B). Also, instead of curved arrows between states to indicate transitions, straight-line arrows can also be used as shown in (B). A variation of table representation would be the omission of the S before the state number:

	x_1	x_2	x_3
1	1	2	2
2	2	2	1, z

Example 6. When several states are involved, any individual state does not always change to the next state, since the transitions are dictated by the problem statement. In Fig. 9-16, for instance, state S_3 is only linked to state S_2. Thus, S_4 cannot be reached directly from S_3. Instead, S_4 state is switched by an x_1 input at S_2.

Note the variation in diagram representation for this illustration. Numbers only appear in the circles for the states, and straight-line arrows are used for transitions between states.

183

Inputs of x_1 at S_2 and S_3 produce a z_1 output, while an input of x_2 at S_2 gives a z_2 output. Since only two inputs and three outputs are present, the table preparation presents no problems:

Present State	Next State and Output x_1	x_2
S_1	2	1
S_2	4, z_1	3, z_2
S_3	2, z_1	3
S_4	1	4

Questions and Problems

1. Define the terms *primaries* and *secondaries* with respect to sequential switching systems.

2. Explain the meaning of the symbols x, z, y, and Y, in sequential switching.

3. To what symbol does the word *excitation* apply and what is its meaning with respect to sequential systems?

4. What memory devices are applicable to sequential switching?

5. Explain the relative functions of signal *pulses* and signal *levels* in sequential switching.

6. What steps are required for synthesis of sequential switching circuitry?

7. Prepare a flow diagram and table for the following problem statement: A sequential switching system is needed requiring three states with two inputs. At S_1 successive x_1 inputs change each state except at S_3, where it is changed with an x_2 input only and produces an output z.

8. Prepare a flow diagram and table for the following problem statement: A sequential switching system has a z_2 level at S_1, a z_1 level at S_2, and a z_3 level at S_3. At S_2 an x_1 returns the state to S_1 and an x_2 to S_3. At S_1 and S_3 only an x_1 input changes state.

9. Prepare a diagram and table for the following: The switching system has three states and two inputs. Each successive x_1 changes state. At S_3 the x_1 input produces a z output. Input x_2 only changes state when state is S_3.

10. Prepare a table for the flow diagram shown in Fig. 9-17.

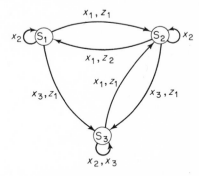

Fig. 9-17. Diagram for Problem 10.

11. Prepare a diagram and table for the following: The switching system is to have four states with two inputs. Successive x_1 and x_2 inputs change state. At S_4 an x_1 switches to S_2 state. At S_2 and S_4 an x_2 input produces a z_1 output. At S_3 and S_4 an x_1 input produces a z_2 output.

12. Prepare a diagram and table for the following: A switching system is to have four states and two inputs. Input x_2 changes state only at S_2, where it switches to S_4 and produces a z_2. At S_4 an x_1 switches back to S_2 with a z_1 output. At S_1 successive x_1 inputs switches to S_2, S_3, S_1, etc. At S_2 an x_1 produces a z_1.

185

10

SEQUENTIAL CIRCUITS

Synthesis

INTRODUCTION

Synthesis is the word used to indicate the combining of individual circuits to form the complete combinational or sequential switching system. In the preceding chapter the x, z, y, Y, symbols were introduced and the utilization of the input x and output z in diagrams and tables was discussed and illustrated. As described in the steps listed on page 176 under "Pulse and Level Signals," however, the problem statement, flow diagram, and table preparation, comprise only the initial procedures for the complete synthesis of sequential circuitry. The additional steps are covered in this chapter and include the reduction of flow diagrams and tables, the assignment of memory states, and the preparation of excitation maps. Thus, this chapter is a continuation of the preceding one and covers the present (y) states of the secondaries and the next states (Y) in association with the principles covered earlier for sequential circuitry.

REDUNDANT FACTORS

As in the synthesis of combinational switching systems, redundant states in sequential circuitry can be eliminated to reduce the total number of states present in the original problem statement. Redun-

186

dancy occurs when two states are *equivalent,* hence one can be omitted (usually the state with the higher numbers). Thus, each occurrence of the higher number is replaced by the lower number. Equivalency is present when one of the following conditions prevails:

1. When two states have the same output level (or output pulse-type signal).
2. When two states (for given inputs) have identical transitions to the same states.

The following examples illustrate each of the conditions which involve redundant states.

Example 1. Assume that a problem statement resulted in the flow diagram shown in Fig. 10-1A. The appropriate flow table is shown

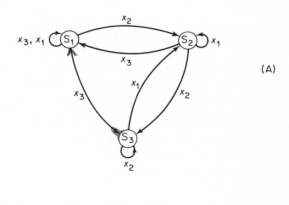

(A)

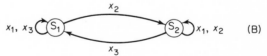

(B)

Fig. 10-1. Reduction. Example 1.

below. (In these initial examples the outputs are not designated.)

Present State	Next State		
	x_1	x_2	x_3
S_1	S_1	S_2	S_1
S_2	S_2	S_3	S_1
S_3	S_2	S_3	S_1

187

For this table, note that for both *present states* S_2 and S_3 sequential x_1, x_2, x_3, inputs provide state transitions which are identical (S_2, S_3, S_1). Because present states S_2 and S_3 are equivalent, one can be eliminated and the switching circuitry reduced. The higher state S_3 is dropped, resulting in a reduction to two present states. For state S_2, however, the x_2 input *next state* S_3 must be changed to S_2 to conform to the reduced two-state condition. The resultant table is as shown below, and the flow diagram as shown in Fig. 10-1B.

Present State	Next State x_1	x_2	x_3
S_1	S_1	S_2	S_1
S_2	S_2	S_2	S_1

Example 2. Redundant states caused by equivalency can be found by careful examination of the flow diagram, but they are more easily ascertained from the table. For the flow diagram in Fig. 10-2A, for

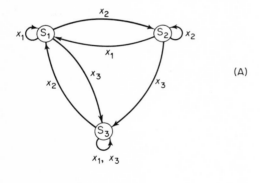

(A)

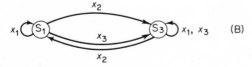

(B)

Fig. 10-2. Reduction. Example 2.

instance, it will be noticed that the three inputs at S_1 produce the same transitional state change they produce at S_2. The accompanying table, however, shows the equivalency immediately without having to follow arrow paths.

188

	x_1	x_2	x_3	
1	1	2	3	
2	1	2	3	(redundant)
3	3	1	3	

When the second state S_2 is eliminated, the third state S_3 replaces S_2 and each 2 is changed to 3 in the reduced table:

	x_1	x_2	x_3
1	1	3	3
3	3	1	3

The two-state table is represented by the reduced flow diagram shown in Fig. 10-2B.

Example 3. Equivalency can, of course, also occur in two-input systems, or those having more than three states. A typical case is the equivalency occurring at S_3 and S_4 in the following table:

	x_1	x_2
1	2	1
2	3	2
3	4	1
4	4	1

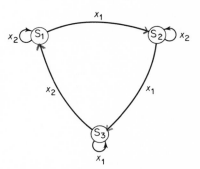

Eliminating the redundant S_4, and changing 4's to 3's, gives us the following reduced table:

	x_1	x_2
1	2	1
2	3	2
3	3	1

Fig. 10-3. Reduction. Example 3.

The flow diagram for the new three-state table is shown in Fig. 10-3 and represents the minimum form for the problem statement associated with the original table.

Example 4. Output (z) designations influence the equivalency of the states. In the following table the transitions to states 1, 2, and 3, are

189

identical for each present state for sequential inputs of x_1, x_2, and x_3. For present states 1 and 2, however, there is no equivalency because the output conditions are not the same:

	x_1	x_2	x_3
1	1	2	3, z
2	1	2, z	3
3	1	2, z	3

Note, however, that for the present states 2 and 3, the same output conditions are present as well as transitions. Hence, the S_3 row can be eliminated and each 3 in the first and second row changed to a 2:

	x_1	x_2	x_3
1	1	2	2, z
2	1	2, z	2

Example 5. When transitional outputs are given, redundancy is present if identical output conditions are listed for two states, as shown in the following table:

Present State	Next State			Next Output		
	x_1	x_2	x_3	x_1	x_2	x_3
1	2	1	1	z_1		z_2
2	1	3	2	z_1	z_2	
3 -	1	3	3	z_1	z_2	

Note that the same output conditions prevail for present states 2 and 3, where z_1 and z_2 occur for x_1 and x_2 inputs. Equivalency of the present states 2 and 3 permits elimination of S_3, and when the 3 in the present-state row S_2 is changed to 2, we have:

Present State	Next State			Next Output		
	x_1	x_2	x_3	x_1	x_2	x_3
1	2	1	1	z_1		z_2
2	1	2	2	z_1	z_2	

Example 6. Present-state levels are shown in the flow diagram of Fig. 10-4A. The table for this switching system shows identical tran-

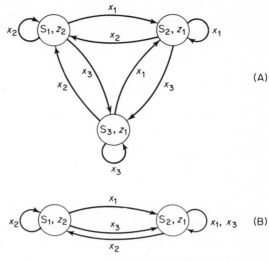

(A)

(B)

Fig. 10-4. Reduction. Example 6.

sitions for sequential inputs of x_1, x_2, and x_3 as well as a z_1 present-state S_2 and S_3 output:

Present State	Next State x_1	x_2	x_3	Present Output
1	2	1	3	z_2
2	2	1	3	z_1
3	2	1	3	z_1

Equivalent states exist for S_2 and S_3 because of the identical transition states (2, 1, 3) and because each present state has the same present output level z_1.

Eliminating the third state S_3 and changing the 3 to a 2 forms the reduced table, and the accompanying flow diagram shown in **Fig. 10-4B**.

	x_1	x_2	x_3	Output
1	2	1	2	z_2
2	2	1	2	z_1

If, in the initial table, present-state S_3 had a transition of 2, 3, 1, instead of 2, 1, 3, the end results would have been the same. Equivalence

191

would still have been present because of the common z_1 output and the third row would still have been eliminated.

Example 7. On occasion, an *optional* or *unspecified* output condition may be encountered as shown in the following table:

Present State	Next State		Present Output
	x_1	x_2	
S_1	S_2	S_1	z_1
S_2	S_1	S_3	
S_3	S_1	S_2	----

For the S_1 state, the present output is indicated as z_1. For the second state (S_2) no output is shown, hence no output terminal is engaged or present for this state. For S_3, however, the dashed line (-----) indicates that an optional or unspecified condition exists, that is, presumably z can be either 0 or 1, and an output line is present for S_3. When such an optional row is present, it is considered to be equivalent in nature or pseudoequivalent to one of the other states, and hence is redundant and can be eliminated. The reduced table thus becomes as follows.

Present State	Next State		Present Output
	x_1	x_2	
S_1	S_2	S_1	z_1
S_2	S_1	S_2	

The optional z output could also have been designated by a solid line (——) instead of -----. In either case it distinguishes the output from that where the space is left blank to show *no* output.

Example 8. The same principle applies to transitional-type tables. In the following, the first state has optional outputs and hence is pseudoequivalent:

Present State	Next State		Next Output	
	x_1	x_2	x_1	x_2
1	2	1	----	----
2	1	3	z_1	
3	2	1		z_2

Thus, we can eliminate the first-state row and renumber the remaining states S_1 and S_2. Also, the remaining 3 is changed to a 2, since the third state no longer exists:

Present State	Next State x_1	Next State x_2	Next Output x_1	Next Output x_2
1	1	2	z_1	
2	2	1		z_2

Example 9. The optional or unspecified factor may be present in a transitional state, as shown in the following table:

	x_1	x_2
1	2, z	----
2	1, z	3
3	1	2, z

Thus, the first-state row can be eliminated as was done in Example 8.

Example 10. When an optional condition occurs with coinciding outputs, the optional section must, of course, be the row that is eliminated. The optional row in the following table is in S_4, and both S_3 and S_4 have the same present-state output level of z_2 as shown:

Present State	Next State x_1	Next State x_2	Present Output
1	2	1	
2	3	2	z_1
3	4	1	z_2
4	----	1	z_2

The reduced table thus is as follows:

	x_1	x_2	Output
1	2	1	
2	3	2	z_1
3	3	1	z_2

Sequential switching systems contain combinational circuitry, and undergo changes of state. Hence, note the results of equivalency when only two states are involved. If they have the same transitions (or the same output), or if one has optional aspects, redundancy is present:

	x_1	x_2	
1	1	2	z
2	1	2	----

When one state is eliminated we have:

	x_1	x_2
1	1	1_z

Now the sequential switching aspects are no longer present and the single-state system is a combinational switching type where $z = x$.

SECONDARY ASSIGNMENTS

The next step in the synthesis of sequential switching systems is the assignment of binary numbers for the state representations of the reduced flow table. The binary numbers indicate flip-flop *states*, with each flip-flop handling two circuit states. Thus, if a sequential switching system has three or four states, two flip-flops are needed. The *on* and *off* states of three flip-flops can represent up to eight circuit states.

The selection of a particular binary-number sequence for a secondary assignment is arbitrary, although one grouping may result in fewer circuits to achieve the same logic end-result system. It is, however, not necessary to try all possible variations, since many yield identical results, or produce only minor differences. When some doubt exists regarding the most economical final switching system, several secondary assignments may be tried, with their particular selection based on experience.

The secondary assignment procedures may be more easily understood by initially using a two-state two-input table as an example:

194

Present State	Next State x_1	x_2
1	2	1
2	1	3
3	3	1_z

As discussed in Chap. 9, page 171, under "Symbols and Terms," the lower-case letter y is used to represent the present state of the secondary (flip-flop memory). Thus, for the *present-state* portion of the foregoing we assign a binary number for each state:

	Present State y_1	y_2
(1)	0	0
(2)	0	1
(3)	1	1

Since we have represented the first state (1) with an 00, we must replace each 1 in the *next-state* columns with 00. Similarly, 2 is now represented by 01, and hence all 2's must be changed to 01 in the *present-state* columns, etc.:

Present State y_1y_2	Next State x_1	x_2
00	01	00
01	00	11
11	11	00_z

Using the same initial table, a different secondary assignment would be a sequence of 00, 11, 10, to represent each of the three states:

y_1y_2	x_1	x_2
00	11	00
11	00	10
10	10	00_z

Similar assignments are made when more states or inputs are present, assigning each similar state the same binary number. If an extra state is shown without input numbers, it indicates that this particular state is unnecessary in the final switching system, and that its usage is optional:

	Initial Flow Table				Secondary Assignments			
	x_1	x_2	x_3		y_1y_2	x_1	x_2	x_3
1	2	1	3		00	01	00	11
2	1	2	3		01	00	01	11
3	3	2_z	3		11	11	01_z	11
					10	----	----	----

As an additional example, two secondary assignments are shown for the following flow table:

	x_1	x_2		y_1y_2	x_1	x_2		y_1y_2	x_1	x_2
1	2	1		00	01	00		00	11	00
2	3	2		01	11	01		11	01	11
3	4	1		11	10	00		01	10	00
4	4	1_z		10	10	00_z		10	10	00_z

TRIGGER EXPRESSION FORMATION

The next step is to obtain the combinational-type expressions from the secondary-assignment table. The methods for doing this vary between the trigger-type flip-flop and the set-reset type. For the flip-flop where only a trigger input is used, the trigger-input pulse will change the state of the flip-flop from either its 0 state to its 1 state, or from its 1 state to the 0 state, as discussed in Chap. 8.

Each term of an expression is obtained by noting if there is a change of state for y_1 or y_2, etc. with respect to input designations x_1, x_2, etc. In the following, for instance, the y_2 state is changed for an x_2 input:

y_1y_2	x_1	x_2
00	00	01

Since the trigger (Tr) is a function of xy, the expression obtained includes the present states of the y's *and* x_2. Thus, the expression is $x_2\bar{y}_1\bar{y}_2$. The y values are shown in negated form since 0's represent them in the table fragment shown above.

196

The input x_1 is involved if it initiates a change in either y_1 or y_2:

y_1y_2	x_1	x_2
10	00	10

Note that under x_1 we now have 00 again, but the first bit differs from that of y_1 which is 1. Hence, the expression becomes $x_1y_1\bar{y}_2$. The $\bar{y}_2$ is now negated, since a 0 appears below it. Since the binary number below x_2 is 10, the same as for the y column, no output expression is obtained for the y_2 and x_2 terms.

Each term obtained from a table with secondary assignments represents an AND function, such as $x_2y_1\bar{y}_2y_3$. Individual terms are combined by the OR function, as $x_2y_1\bar{y}_2y_3 + x_1\bar{y}_1y_2$.

As an example of the complete synthesis process, assume that we require a switching system with two states where successive x_2 and x_1 inputs change states. An output z is to be obtained when an x_2 input changes S_1 to S_2. The flow diagram

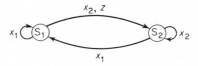

Fig. 10-5. Flow diagram reference.

for this stated problem is shown in Fig. 10-5. From this we obtain the initial flow table:

	x_1	x_2
1	1	2_z
2	1	2

Assigning 00 to 1 and 01 to the second state, we obtain the following table. Note that in the first state there is a change for y_2 for an x_2 input, producing $x_2\bar{y}_1\bar{y}_2$. In the second state, y_2 is again changed, but now by x_1 to form $x_1\bar{y}_1y_2$:

y_1y_2	x_1	x_2	
00	00	01_z	
			$\ldots\bar{y}_1\bar{y}_2x_2$
01	00	01	
			$\ldots\bar{y}_1y_2x_1$

$$\text{Tr} = \bar{y}_1\bar{y}_2x_2 + \bar{y}_1y_2x_1$$

From the table with the secondary assignment, it will be noted that the state of flip-flop $\bar{y}_1$ does not change, indicating that only a single flip-flop stage is required for the stated conditions. Thus a single flip-flop with the $\bar{y}$ present-state output can be used. Now the trigger expression becomes:

$$\text{Tr} = x_1\bar{y} + x_2\bar{y}$$

The z output must occur during the first state, that is, when $\bar{y}$ is in negated form and an x_2 input is used. Thus, the expression for the output is:

$$z = x_2\bar{y}.$$

The expression for the trigger shows that two AND circuits are needed plus one OR switch. Since the z output expression is the same as one of the Tr members, the same AND circuit can be used for both. Initially the symbols for the flip-flop, OR, and AND circuits are drawn. In the first state only x_2 input triggers and it requires an accompanying $\bar{y}$. Thus, as shown in Fig. 10-6A, the output $\bar{y}$ from the flip-flop is

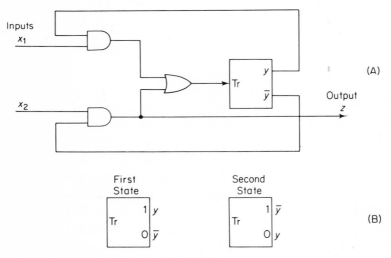

Fig. 10-6. Switching circuitry for Fig. 10-5.

applied to the input of the AND circuit with the x_2 input. In the first state, however, an x_1 input has no effect. Thus the other output of the flip-flop (y) is applied to the AND circuit having an x_1 input. From the

198

expression for the trigger it is seen that an x_1 input is not effective unless accompanied by a $\bar{y}$, hence an x_1 input will not trigger the flip-flop while it is in the 0 state.

If the flip-flop is triggered by an x_2 input it changes to the second state, with y outputs as shown in Fig. 10-6B. Note that now a y (unnegated) is applied to the x_2 AND circuit which prevents an x_2 input from triggering during the second state of the flip-flop. For an x_2 input we must have the accompanying $\bar{y}$ (negated). For the x_1 input, however, we now have the necessary $\bar{y}$ input to permit triggering. Thus, only x_1 can trigger when the flip-flop is in its second state.

Because an output is obtained from the x_2 input when accompanied by $\bar{y}$, the output z is obtained from the output of the x_2 AND circuit as shown.

If we had used a different secondary assignment, the resultant logic would have been the same, though the circuit arrangement would be slightly different. Thus, if we had used 11 for the first state, and 10 for the second state, the table appears as:

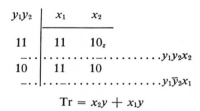

$$\text{Tr} = x_2 y + x_1 y$$

Now our end product results in the switching system shown in Fig. 10-7. Input x_2 still triggers when the flip-flop is in the first state,

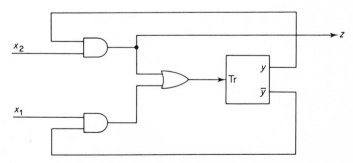

Fig. 10-7. Circuitry for different secondary assignment.

199

except that now it must be accompanied by a y instead of the negated form, as shown in the expression for Tr obtained from the table.

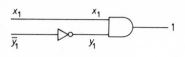

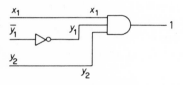

Fig. 10-8. Logic input functions.

Identical AND-gate symbols were shown for Fig. 10-6 and Fig. 10-7, though one type AND switch called for a negated input ($\bar{y}$) with an unnegated x. It is assumed that appropriate NOT circuits will be utilized to satisfy the logical requirements. Logic input functions could take the forms shown in Fig. 10-8A and B, where the NOT (*inverters*) assure coinciding inputs to open the AND gate circuitry. Thus, if an x_1 input requires an accompanying $\bar{y}$, the inverter produces the required x and y inputs at the AND circuit. No output is produced if the input is $\bar{x}_1\bar{y}$ or x_1y.

MULTISTATE EXPRESSIONS

In the foregoing examples for obtaining expressions, only a single flip-flop (with two states) was involved. With three or four states, however, two flip-flop circuits are required, hence there will be two trigger inputs, Tr_1 and Tr_2. Thus, an individual expression must be obtained for each trigger input. This is exemplified in the following table:

Present State	Next State	
	x_1	x_2
1	1	2
2	3	2
3	4	3
4	4	1_z

Using the Gray-code sequence for the secondary assignments produces the following table and expressions. Any expression involving y_1 is listed under flip-flop 1 (FF1) and involves trigger Tr_1. Terms

200

produced from the y_2 section are listed under flip-flop 2 (FF2) and relate to trigger Tr_2.

y_1y_2	x_1	x_2	FF1	FF2
00	00	01		
$-..$				$x_2\bar{y}_1\bar{y}_2$
01	11	01		
$-...$			$x_1\bar{y}_1y_2$	
11	10	11		
$-..$				$x_1y_1y_2$
10	10	00_z		
$-...$			$x_2y_1\bar{y}_2$	

Gathering terms, we have, for FF1

$$\text{Tr}_1 = x_1\bar{y}_1\bar{y}_2 + x_2y_1\bar{y}_2.$$

(handwritten: $x_1\bar{y}_1\bar{y}_2$)

For FF2 we obtain

$$\text{Tr}_2 = x_1y_1y_2 + x_2\bar{y}_1\bar{y}_2.$$

For the output z, we require an x_2 input at the time y_1 is 1 and y_2 is 0, hence

$$z = x_2y_1\bar{y}_2.$$

Inspection of the Tr_1 and Tr_2 expressions shows that we require two AND switches and one OR switch for each trigger flip-flop, for a total of four AND circuits and two OR circuits as shown in Fig. 10-9. An additional AND circuit is required for the z output.

For ease of analysis, assume that each AND circuit opens only when coinciding x and y signals are present. Any negated term such as $\bar{y}_1$ or $\bar{y}_2$ appearing *directly* at the AND-gate inputs causes a closed-gate condition and no output is obtained. NOT circuits are shown for convenience in understanding operation, though these are usually omitted in the logic draft for initial illustration of the complete sequential logic system.

Note from the secondary-assignment table that when the system is in the first state only an x_2 input is active, since it triggers the second flip-flop. In the first state, however, x_1 has no triggering effect. For x_2 to trigger the second flip-flop we must have $x_2\bar{y}_1\bar{y}_2$ as shown in the expression for Tr_2. Note that this condition is satisfied for AND gate

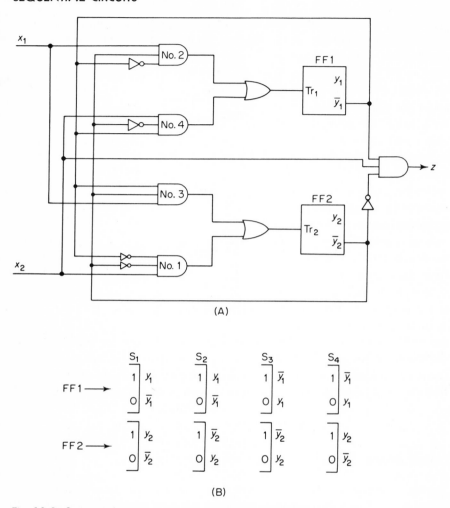

Fig. 10-9. Sequential system for four-state, two-input T flip-flops.

No. 1 in Fig. 10-9, since the $\bar{y}$ output lines from each flip-flop are applied to the input leads of this AND gate, in addition to x_2. The two NOT circuits invert the $\bar{y}$ signals to provide for coincidence directly at the AND circuit inputs. No other AND circuit opens during the first state, because a $\bar{y}$ appears at one of its input lines without inversion.

When in the second state, as shown under S_2 in Fig. 10-9B, FF2 produces an unnegated y from its output terminal. Now only an x_1 input triggers, because at AND gate No. 2 the necessary $x_1\bar{y}_1y_2$ signals appear. The y signal is inverted, as shown, and with the x_1 input and the y_2 obtained from the second flip-flop, an open-gate AND function produces an output to trigger flip-flop FF1. During state S_2 no other AND gate provides an output. (At AND gate No. 1, for instance, the y_2 input is inverted and appears as $\bar{y}_2$, thus preventing the gate from opening.)

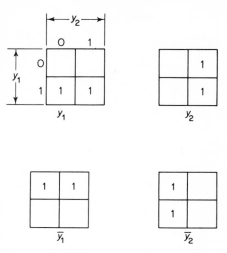

Fig. 10-10. Logic maps for y_1, y_2.

Similar analysis for S_3 and S_4 states will show that the proper switching sequence prevails as indicated in the logic tables. In the fourth state a y_1 is obtained from the output of FF1 and a $\bar{y}_2$ from FF2. Only at this time is a z output produced for an x_2 input. The $\bar{y}_2$ is inverted by the NOT circuit and provides the necessary coincidence for opening the output AND circuit. It is also at this time that an x_2 input will cause coincidence at AND gate No. 4 to retrigger FF1 to its original first state and hence the S1 condition of the complete sequential system.

EXCITATION MAPS

Another method for obtaining expressions from secondary assignment tables is to employ excitation maps. These are similar to those shown earlier in Fig. 2-4, except that y_1 is substituted for A and y_2 for B as shown in Fig. 10-10. One advantage is that minimum expression forms are obtained without having to use Boolean algebra for eliminating redundancies. This factor will be illustrated later in this chapter.

The use of excitation maps for obtaining expressions is illustrated in Fig. 10-11 for the four-state sequential system discussed in the previous section. The upper two maps are for the first flip-flop and the lower two for the second. Inputs x_1 and x_2 are represented by separate maps for each flip-flop, as shown.

203

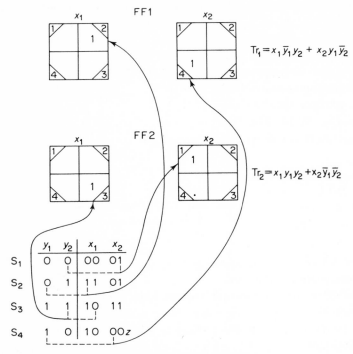

$$Tr_1 = x_1 \bar{y}_1 y_2 + x_2 y_1 \bar{y}_2$$

$$Tr_2 = x_1 y_1 y_2 + x_2 \bar{y}_1 \bar{y}_2$$

Fig. 10-11. Excitation maps for four-state problem.

The numbers in each map represent the successive states. Thus, for the first state shown in the secondary-assignment table, y_2 changes state for an x_2 input, hence a 1 is placed in the first-state block of the lower right map. For the second state shown in the table, y_1 changes state for an x_1 input. This requires placing a 1 in the second block of the upper left map to represent the second state for the first flip-flop with an x_1 input.

The remaining map entries are entered in similar fashion, until every y_1 and y_2 change-of-state has been notated. Now the maps are read for the terms they represent, keeping in mind that for an AND function any square not coincident is left blank as described in Chap. 2. From reference to the values shown in Fig. 10-10 the upper left map in Fig. 10-11 shows coincidence for $\bar{y}_1$ and y_2, and hence yields the $x_1 \bar{y}_1 y_2$ member of the Tr_1 expression. The second member is obtained from the upper right map where the 1 in the fourth block indicates a coincident condition for y_1 and $\bar{y}_2$. Reference to Fig. 10-10 will also indicate

204

the combinational members of the expression for Tr_2 for the lower two maps.

As mentioned earlier, mapping procedures produce the minimum form
of the expression and are convenient
where lengthy expressions are involved.
As an example, consider the flow diagram shown in Fig. 10-12. The flow
table for this diagram is as follows:

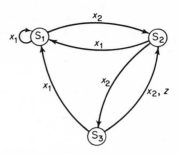

	x_1	x_2
1	1	2
2	1	3
3	1	2, z
4	----	----

Fig. 10-12. Flow diagram for
three-state problem.

Because two flip-flop stages are required, four states are available,
with the fourth designated as optional.

If we use the Gray-code numbering sequence for the secondary
assignment, we obtain the following table. Note that there are six
changes-of-state involved, three of which occur in the S_3 row (two for
x_1 and one for x_2 inputs).

y_1y_2	x_1	x_2	FF1 (Tr_1)	FF2 (Tr_2)
00	00	01 =		$x_2\bar{y}_1\bar{y}_2$
01	00	11 =	$x_2\bar{y}_1y_2$	$x_1\bar{y}_1y_2$
11	00	01$_z$ =	$x_1y_1y_2$	$x_1y_1y_2$
10	----	----	$x_2y_1y_2$	

Gathering terms we now have:

$$Tr_1 = x_1y_1y_2 + x_2y_1y_2 + x_2\bar{y}_1y_2$$
$$Tr_2 = x_2\bar{y}_1\bar{y}_2 + x_1\bar{y}_1y_2 + x_1y_1y_2$$

By Boolean algebra principles we can reduce the x_2 members in the
Tr_1 expression:

$$x_2y_2(y_1 + \bar{y}_1) = x_2y_2(1 + 0) = x_2y_2.$$

Thus, the Tr_1 expression becomes

$$Tr_1 = x_1y_1y_2 + x_2y_2.$$

205

Similar treatment of the x_1 members of the Tr_2 expression produces:

$$Tr_2 = x_1y_2 + x_2\bar{y}_1\bar{y}_2.$$

Hence, by term reduction, we have reduced each trigger expression by one member. With the excitation map entries, however, the minimum forms are immediately apparent as shown in Fig. 10-13. For the

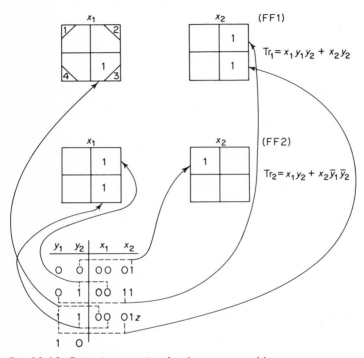

Fig. 10-13. Excitation mapping for three-state problem.

x_2 map at the upper right the two 1's indicate a y_2 literal as shown in Fig. 10-10. Similarly, for the lower left map a y_2 is procured.

The output z occurs during the third state for an x_2 input as shown in the tables, where $y_1y_2 = 11$. Thus, the output expression $z_2y_1y_2$ is valid, except that the y_2 is redundant, because the z output can be identified with y_1 alone since it only occurs when y_1 is in the 1 state. Thus, $z = x_2y_1$.

Had the fourth state not been optional, we could not have eliminated the y_2 literal, because we would have obtained an output again in the fourth state where y_1 also equals 1. With the fourth state optional, however, the next state change after the third would be the first stage, hence $z = x_2 y_1$ suffices.

The diagram for this three-state problem is shown in Fig. 10-14.

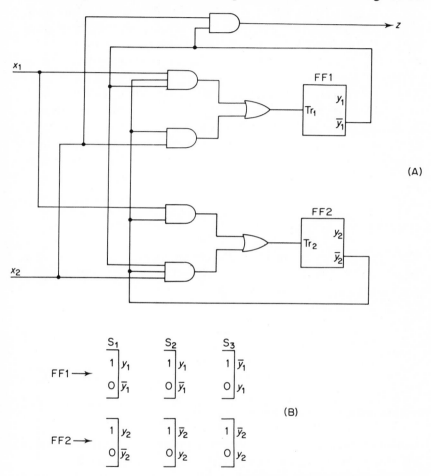

Fig. 10-14. Diagram for three-state problem.

Inverters are not shown, but presumed to be present in the final design, as shown earlier in Figs. 10-8 and 10-9. From the tables it will be noted that when the system is in its first state only an x_2 input changes the state. Since FF2 is involved, the x_2 tripping logic is $x_2 \bar{y}_1 \bar{y}_2$ which is applied to the lower AND circuit for tripping Tr_2. When the system

207

is in the second state as shown in Fig. 10-14B, FF2 output is a y_2, which will permit the tripping of Tr_1 with an input x_2y_2 or the tripping of Tr_2 with an input of x_1y_2, using the two-input AND switches shown in Fig. 10-14.

When in the third state, unnegated y's are available from each output terminal of the flip-flops, as shown in Fig. 10-14B. Now an x_1 input will trigger both Tr_1 and Tr_2 with a change of state back to S_1. With an x_2 input, however, only the Tr_1 flip-flop changes state, as shown by the tables.

It must be remembered that the system shown in Fig. 10-14 represents an individual secondary assignment. Other secondary assignments may be tried, if it appears that circuit reduction is a possibility. In logic-circuit design practices the aim is for a minimum of switching circuitry to keep the manufacturing cost as low as possible. Later examples in this chapter will emphasize this factor to a greater extent.

MULTIPLE-INPUT MAPPING

When three inputs are required, three maps will have to be used for each horizontal row representing a flip-flop. Assume, for instance, that we have the following flow table:

	x_1	x_2	x_3
1	2	1	3_{z_1}
2	1	2_{z_2}	3
3	3	2	3

Selecting an arbitrary secondary assignment we obtain the following:

y_1y_2	x_1	x_2	x_3
00	01	00	10_{z_1}
01	00	01_{z_2}	10
10	10	01	10

The excitation mapping is shown in Fig. 10-15. (Note the change in the number sequence around the block maps. This is necessary to conform to the secondary assignment selected. Had we again used the Gray-code ordering, the numbering would have been as shown in Fig. 10-13. Literals shown in Fig. 10-10, however, still apply.)

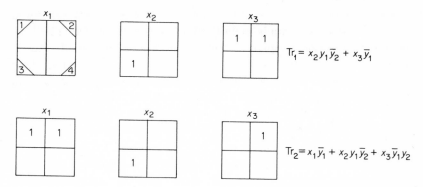

$Tr_1 = x_2 y_1 \bar{y}_2 + x_3 \bar{y}_1$

$Tr_2 = x_1 \bar{y}_1 + x_2 y_1 \bar{y}_2 + x_3 \bar{y}_1 y_2$

Fig. 10-15. Three-input maps.

For the Tr_1 output a three-input AND switch as well as a two-input type are required, plus an OR circuit. For Tr_2 we require three AND switches, but can use a single three-input OR for feeding the Tr_2 input. The two outputs are $z_1 = x_3 \bar{y}_1 \bar{y}_2$ and $z_2 = x_2 \bar{y}_1 y_2$. Separate AND circuits are used as shown in Fig. 10-16 for the complete system logic circuitry.

As explained in the *Introduction* to Chap 7. there are 2^n functions

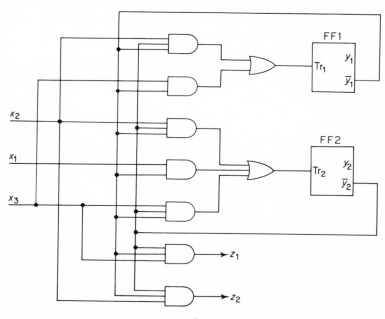

Fig. 10-16. Three-input logic circuitry.

209

of an n variable having two states (1 and 0). Thus, if three flip-flop stages are used, there can be eight combinations of the y literal ($2^3 = 8$), as shown in earlier chapters for truth tables involving A, B, and C. Thus, if the sequential system has from five to eight states, three flip-flop stages would be used, and the y variables would be y_1, y_2, and y_3. This is exemplified in the flow diagram shown in Fig. 10-17.

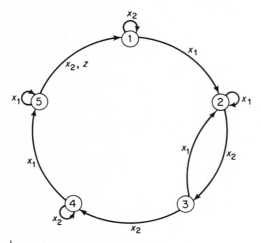

Fig. 10-17. Five-state flow diagram.

The flow table for this diagram shows no redundancies, hence the flow diagram cannot be reduced in states.

	x_1	x_2
1	2	1
2	2	3
3	2	4
4	5	4
5	5	1_z

For mapping, the three-variable types discussed in Chap. 7 are used. Using the Gray code for secondary assignments we obtain the table shown in Fig. 10-18. The arrows show the derivation of the map entries. Instead of following numbered squares, the binary bits under the three y literals are transferred to the appropriate map. Thus, in the first state, y_2 undergoes a change of state for an x_1 input. Thus, the 000 is carried to the x_1 map for Tr_2, as shown, and a 1 placed in the

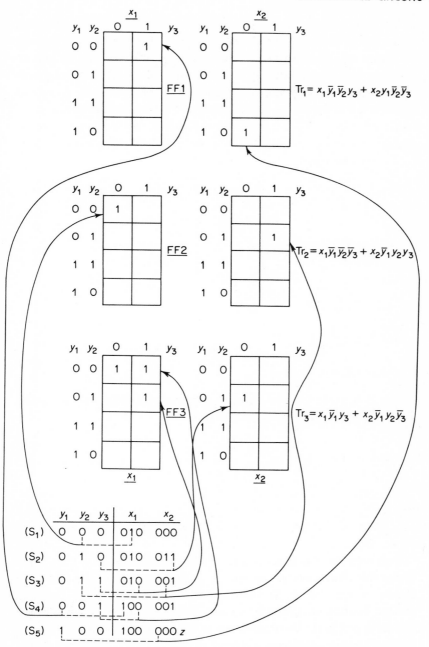

Fig. 10-18. Five-state mapping.

square representative of $\bar{y}_1\bar{y}_2\bar{y}_3$. For the second state, y_3 is triggered by an x_2 input. Hence, the 010 is carried to the x_2 map for the third flip-flop. Note that there are seven changes of state for various inputs, but the map readings only yield six, illustrating the expression-reduction factors relating to map simplification.

The trigger expressions indicate two AND circuits each feeding an OR circuit. Each AND circuit has four inputs except for the x_1 input that triggers Tr_3. The resultant logic circuitry is shown in Fig. 10-19.

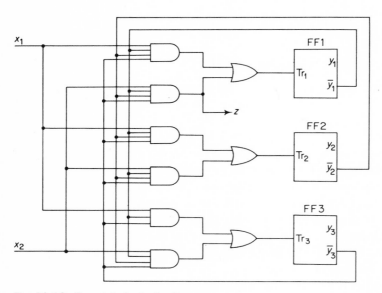

Fig. 10-19. Five-state logic circuitry.

The x_2 input that trips the first flip-flop produces an output from the second AND switch only when S_5 prevails, at which time the output values from the flip-flop stages are $y_1\bar{y}_2\bar{y}_3$. Thus the z output can be obtained from this AND switch because the z expression is also $x_2y_1\bar{y}_2\bar{y}_3$.

Four two-state variables can have 16 combinations ($2^4 = 16$), hence where 9 to 16 states are involved, the Karnaugh maps described in Chap. 7 can be used. These accommodate the necessary y_1, y_2, y_3, and y_4 variables as shown in Fig. 10-20. This could represent the two inputs x_1 and x_2 for up to 16 states. For the maps shown, FF1 trigger reads $x_1y_1y_2\bar{y}_3y_4 + x_2\bar{y}_1y_2y_4$. If three x inputs are involved, three such maps would have to occupy a single row for a specific flip-flop stage.

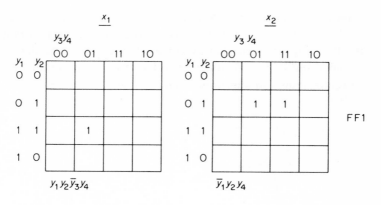

$$Tr_1 = x_1 y_1 y_2 \bar{y}_3 y_4 + x_2 \bar{y}_1 y_2 y_4$$

Fig. 10-20. Karnaugh excitation maps.

S-R SYNTHESIS

Flip-flop stages using set (S) and reset (R) inputs can also be used instead of the trigger (Tr) input types, or combinations of these may be employed. On occasion it may be advantageous to use the S-R types or combinations for obtaining a minimum of switching circuits in the final form. To illustrate this factor, the Tr flip-flop will be used initially. Using the same flow table, the set-reset type will be analyzed for comparison. The difference in the end-result using another secondary assignment will also be shown.

The following is the flow table for the problem to be solved:

	x_1	x_2
1	2	1
2	2	3
3	1	3_z

Using 00, 01, 10 for the secondary assignment, we obtain the following table and trigger expressions:

$y_1 y_2$	x_1	x_2
00	01	00
01	01	10
10	00	10_z

$$Tr_1 = x_1 y_1 \bar{y}_2 + x_2 \bar{y}_1 y_2$$
$$Tr_2 = x_1 \bar{y}_1 \bar{y}_2 + x_2 \bar{y}_1 y_2$$
$$z = x_2 y_1 \bar{y}_2$$

213

Note that the $x_2\bar{y}_1y_2$ is common to both Tr_1 and Tr_2, hence a single AND switch can be used to trigger either state for an x_2 input. The resultant logic circuitry, including the AND switch for the z output, includes four AND circuits and two OR circuits, for a total of six logic gates, as shown in Fig. 10-21. The changes of each state are shown in (B).

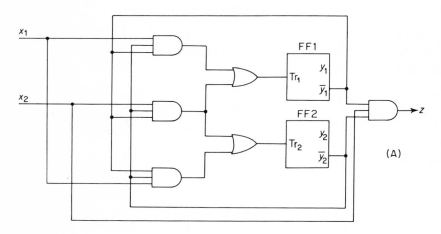

Fig. 10-21. Common Tr_1, Tr_2 AND switch.

For comparison purposes with the S-R flip-flops with the same assignment table, we must separate the individual x_1 and x_2 equations to represent the *set* conditions (from 0 to 1) or the *reset* (1 to 0). Thus, for a change of y_1 from 0 to 1, we obtain an S_1 term. If the change of y_1 is from 1 to 0, it represents a reset R_1. Similarly, for a change of 0 to 1 for y_2 we obtain an S_2 term. An R_2 term is obtained for a change from 1 to 0 for y_2. When obtaining the terms directly from the table, the S and R terms are listed below the FF1 and FF2 headings, as shown in the following table.

y_1y_2	x_1	x_2	FF1	FF2
00	01	00		$S_2 = x_1\bar{y}_1\bar{y}_2$
01	01	10	$S_1 = x_2\bar{y}_1y_2$	$R_2 = x_2\bar{y}_1y_2$
10	00	10_z	$R_1 = x_1y_1\bar{y}_2$	

The output is the same as for the trigger flip-flop circuitry: $z = x_1y_1\bar{y}_2$. Now, with the S-R circuitry, only five logic gates are required, as shown in Fig. 10-22.

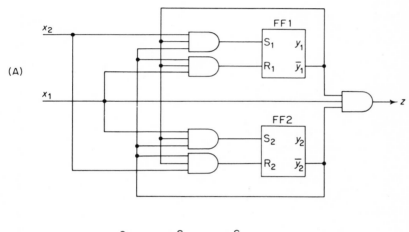

(A)

(B)

Fig. 10-22. Set and reset switching.

Excitation maps can also be employed for obtaining the individual S and R expression members. As shown in Fig. 10-23, if a change from 1 to 0 at S_3 occurs for y_1 for an x_1 input, it indicates a reset function, hence an R is placed in the third square of the x_1 map for FF1. Simi-

215

larly, if y_1 changes from 0 to 1 for state 2, with an x_2, it indicates a set condition and requires an S in the second square of the x_2 map of FF1.

Similar mapping is undertaken for any y_2 change. If an x_1 input changes y_2 from 0 to 1 during the first state, an S is placed in the first square of the x_1 map for FF2. If an x_2 input at state 2 changes y_2 from 1 to 0, an R is placed in the second square of the x_2 map for FF2 as shown in Fig. 10-23. The individual map entries are then read (Fig.

FF1

$S_1 = x_2 \bar{y}_1 y_2$

$R_1 = x_1 y \, \bar{y}_2$

FF2

$S_2 = x_1 \bar{y}_1 \bar{y}_2$

$R_2 = x_2 \bar{y}_1 y_2$

Fig. 10-23. S-R excitation maps.

10-10) and the S and R terms obtained. (Note the numbering sequence used to conform to the particular secondary assignment employed.)

The complete system shown in Fig. 10-22 should not be assumed to be in minimum form unless several other secondary assignments have been tried. A saving of an additional logical gate can be made, for instance, with the following secondary assignment:

y_1y_2	x_1	x_2	
00	10	00	$S_1 = x_2 y_1 \bar{y}_2$
10	10	11	$S_2 = x_1 \bar{y}_1 \bar{y}_2$
11	00	11,	R_1 and $R_2 = x_1 y_1 y_2$

Since the reset expressions are the same for both flip-flop states, a stage is saved. Thus, only four logical gates are needed as shown in Fig. 10-24. The x_1 input could have been applied directly to the R input of FF2, because it would be ineffectual during the first two states, where FF2 is already in the reset (0) state. It is only in state 3 that x_1

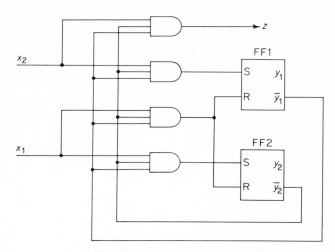

Fig. 10-24. Reduced circuitry for S-R switching.

resets FF2. Since no logic-gate saving results, however, both R inputs can be obtained from the AND circuit as shown.

S-R Reduction Factors

When S-R flip-flop stages are used, logical gates can often be eliminated because of the inherent limited functions of set and reset. If several 0's exist in sequence for states of y_1, for instance, an R input has no effect until preceded by an S to set the state to 1. Similarly, a succession of 1's for a flip-flop indicates a set condition and hence an S input is ineffective. Often an inspection of the secondary-assignment table based on logical considerations will reveal areas where simplification is possible. As an example of this, consider the following table with its resultant S and R members:

y_1y_2	x_1	x_2	FF1	FF2
00	00	11	$S_1 = x_2\bar{y}_1\bar{y}_2$	$S_2 = x_2\bar{y}_1\bar{y}_2$
11	01	11	$R_1 = x_1y_1y_2$	
01	00_z	11	$S_1 = x_2\bar{y}_1y_2$	$R_2 = x_1\bar{y}_1y_2$

Since there are dual x_2 inputs for S_1 we have:

$$S_1 = x_2\bar{y}_1\bar{y}_2 + x_2\bar{y}_1y_2 = x_2\bar{y}_1$$

217

Since, however, the S function can *set* only, and not trigger, the literal contingent $\bar{y}_1$ is not necessary since obviously y_1 has to be in the 0 state to set. Hence, $S_1 = x_2$ only. Note that S_2 is identical to S_1 in the derived expressions from the table, hence the same factors apply and $S_2 = x_2$.

From the table R_1 is indicated as $x_1 y_1 y_2$. An inspection of the state sequence of the table, however, shows that y_1 is reset only once: at the second state with an x_1. Obviously it must be in the 1 state to be reset, hence the contingent y_1 literal is meaningless here. The second literal y_2 is also of no import, since there are no other contingencies where R_1 must have a y_2 to reset. Hence, $R_1 = x_1$.

For R_2 the table yields $x_1 \bar{y}_1 y_2$. Obviously y_2 must be in the 1 state to reset, hence the y_2 contingency of the expression can be eliminated. Note, however, that at state 2 an x_1 *does not change the state* of y_2, while at state 3 an x_1 does. Since x_1 does not change the state of y_2 each time, it is contingent on some *state* factor. From the table it can be seen

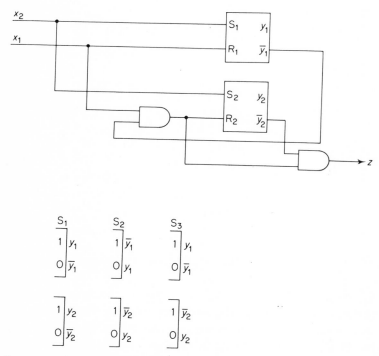

Fig. 10-25. Simplified S-R switching.

that the reset function only applies to y_2 at the time y_1 is 0 ($\bar{y}_1$). Hence, the contingent $\bar{y}_1$ must be retained with the x_1 for R_2: $R_2 = x_1\bar{y}_1$. Now our new S and R listings are:

$$S_1 = x_2$$

$$R_1 = x_1$$

$$S_2 = x_2$$

$$R_2 = x_1\bar{y}_1.$$

These simplified terms result in a considerable reduction in the number of logical circuits necessary to perform the required switching. With the original terms we would have required four logic gates (with the x_2 members for S_1 combined into one), plus an AND switch for the output. For the simplified terms x_2 input can be applied to both S inputs of the flip-flop stages. Similarly, the R_1 input can be obtained directly from the x_1 line. The resultant circuitry is shown in Fig. 10-25, where only two logical gates are used.

As shown in Fig. 10-26, the mapping process would have resulted in

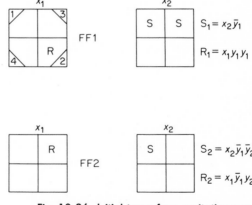

Fig. 10-26. Initial terms from excitation map.

the single S_1 expression. Without a logical analysis of the secondary-assignment table, however, no logical-gate circuit reduction would have been indicated. Again note the numbering sequence within the squares of the maps. This particular sequence is necessary for the secondary assignment selected.

219

The following table is another example of term simplification by logical analysis.

y_1y_2	x_1	x_2
00	01	00
01	11	00
11	11	00_z

The mapping for this table, plus the resultant S and R terms are shown in Fig. 10-27. Four AND switches are indicated, plus another

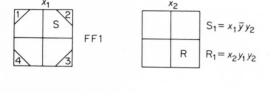

$S_1 = x_1 \bar{y} y_2$

$R_1 = x_2 y_1 y_2$

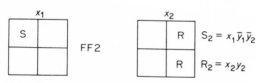

$S_2 = x_1 \bar{y}_1 \bar{y}_2$

$R_2 = x_2 y_2$

Fig. 10-27. S-R maps for problem.

one for the z output. For S_1 we have $x_1\bar{y}_1y_2$. Input x_1 does not set y_1 at the first state, but does so at the second state. Hence, S_1 is contingent on the presence of y_2. Since, however, y_1 can't be set unless it is $\bar{y}_1$, this literal is redundant in the S_1 expression. Thus, $S_1 = x_1y_2$ only.

Note that when either or both y's are in the 1 state, an x_2 will cause it to reset, regardless of the state of the other flip-flop. Hence, R_1 and R_2 require only an x_2 to reset to 0. Note also that y_2 is 0 only during the first state. At that time it is tripped to the set (1) state by x_1. Since y_2 remains in the set position for the second and third system states, x_1 has no additional affect, since it can't reset. Thus, $S_2 = x_1$ only. Output $z = x_2y_1$.

The simplified terms may now be listed as:

$$S_1 = x_1y_2$$

$$R_1 = x_2$$

$$S_2 = x_1$$
$$R_2 = x_2.$$

The switching system based on these S and R terms is shown in Fig. 10-28. Note that again only two AND switches are needed to perform the required functions indicated in the table.

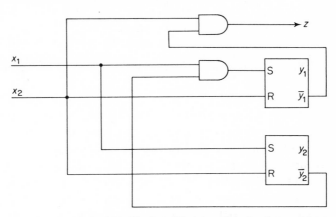

Fig. 10-28. Simplified switching for S-R problem.

In some instances oversimplification may result in producing invalid term combinations. Such a case would be where both S_1 and R_1 are x_1. In the reset-set flip-flops it is presumed that the R and S input will never be applied simultaneously, since one function negates the other. Thus, a single x_1 line cannot be applied to both the S_1 and R_1 inputs to a flip-flop. An x_1 can, however, be applied simultaneously to S_1 and R_2 inputs, since separate flip-flop stages are involved, and the setting of one can also cause the resetting of the other by a single x_1 input. For a single flip-flop stage, however, either the S_1 or R_1 input must have a contingent accompanying literal if a common x_1 or x_2 is involved. The x_1y_1 type term will provide for an AND switch input and prevent the setting of the flip-flop during the time an x_1 reset is present (or vice versa).

COMBINED S-R AND TR

The S-R flip-flop stages can be combined with the Tr types, and in some instances there may be a reduction in logical circuitry. When

221

one flip-flop is an S-R type and the other a Tr type, it will be necessary to list the individual S and R terms for one and the Tr term for the other as shown in the following secondary-assignment table:

y_1y_2	x_1	x_2	Tr_1	S_2	R_2
00	01	00		$x_1\bar{y}_1\bar{y}_2$	
01	01	11	$x_2\bar{y}_1y_2$		
11	01_z	00	$x_1y_1y_2$		$x_2y_1y_2$
10	----	----	$x_2y_1y_2$		

The two x_2 terms for Tr_1 can be combined in a single term:

$$Tr_1 = x_1y_1y_2 + x_2y_2.$$

For an x_1 input y_2 (FF2) sets during the first state. Since S_2 cannot reset and is not contingent on any other states, $S_2 = x_1$ only. The reset R_2, however, does not occur at the second state where y_2 is 1, but occurs only during the third state. Since this is the only time $y_1 = 1$ (except for the optional fourth state which is not used), R_2 is contingent on y_1, hence $R_2 = x_2y_1$. [The y_2 literal in the R_2 term is redundant because R_2 can only set when $y_2 = 1$ and R_2 has no effect when $\bar{y}_2$ (0) is present because it is already in the reset (0) state.]

The output z occurs only for x_1 when $y = 1$, hence $z = x_1y_1$. Thus the output can be obtained from the AND circuit having conditions $x_1y_1y_2$. It is only at the third state that these conditions prevail and produce an output from this AND circuit. The complete system is shown in Fig. 10-29. Again, this may not be the most simplified form

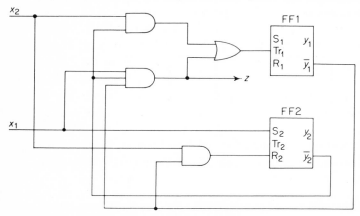

Fig. 10-29. Combined Tr and S-R switching.

obtainable for the original problem statement. To explore fully the possible simplification of the final system, other secondary-assignment tables should be tried.

Questions and Problems

1. Prepare new tables for the following with all redundancies eliminated:

	x_1	x_2	x_3
1	2	1	3
2	2	1	3
3	3	2	1_z

	x_1	x_2
1	2	1
2	2	3_z
3	1	4
4	1	4

	x_1	x_2	
1	2	1	z_2
2	2	3	z_1
3	1	3	z_3
4	4	1	z_3

2. Replace each of the following with redundancies eliminated:

	x_1	x_2	x_3
1	3	1_z	2
2	3	1	2_z
3	3	1	2_z
4	3	1	4_z

	x_1	x_2	x_3	
1	3	1	2	
2	2	1	3	z
3	3	4	1	
4	4	3	1	----

3. Make secondary assignments for the following two tables, using Gray-code ordering:

	x_1	x_2
1	2	1
2	2	3_z
3	3	1

	x_1	x_2	x_3
1	1	2	1
2	2	1	3
3	3	1	4
4	4	3	1_z

4. For the three-input four-state flow table in Prob. 3, prepare three additional secondary assignment tables in addition to the Gray-code ordering.

5. Prepare a secondary-assignment table for the following, using Gray-code ordering:

	x_1	x_2	x_3
1	1	2	1
2	3	2_z	1
3	3	1	4
4	----	----	----

6. Using Gray-code sequence for states, prepare a secondary-assignment table for the one shown below and obtain Tr_1, Tr_2, and z expressions.

	x_1	x_2
1	2	1
2	2	3
3	3	4
4	4	1_z

7. Assign your own secondary values to the table below and solve for Tr_1, Tr_2, and z.

	x_1	x_2	x_3
1	2	1	1
2	2	3	1
3	3	1	3_z

8. Prepare a logical-switching sequential circuit diagram for the secondary-assignment table prepared for Prob. 6.

9. From the following table, obtain minimum forms for S_1, R_1, S_2, R_2, and z:

y_1y_2	x_1	x_2
00	00	01
01	11	01
11	11_z	10
10	00	10

10. Prepare a complete logical-circuit diagram for the S, R and z terms obtained in Prob. 9.

11. Prepare a flow diagram, a flow table, and a secondary-assignment table for the problem:

> The system has three states and two inputs. When in the first and third states, x_2 does not change state. When in the second or third state, x_1 returns the system to the first state. Input x_1 changes S_1 to S_2, and x_2 trips S_2 to S_3 at which time an output z is obtained.

12. Use excitation maps for obtaining Tr_1, Tr_2, and z terms from the assignment table prepared for Prob. 11.

13. From the following table, obtain S_1, R_1, S_2, R_2, and z values. Check results obtained by using excitation maps, and obtain minimum expression forms.

y_1y_2	x_1	x_2	x_3
00	00	01	00
01	00	11	01
11	01$_z$	11	01

14. Prepare a complete logical-circuit diagram for the S, R, and z values obtained for Prob. 13.

15. Repeat Probs. 13 and 14 for two other secondary assignments.

16. What type excitation maps must be used for a sequential system with a six-state switching sequence?

17. What type excitation maps must be used for a sequential system having 12 states?

18. In mapping a ten-state problem, the following Tr values were obtained: $Tr_1 = x_1y_1y_2\bar{y}_3 + x_2y_1\bar{y}_4$ and $Tr_2 = x_3y_1\bar{y}_2\bar{y}_3 + x_2\bar{y}_1y_2\bar{y}_4$. What type maps were used and what were the minimum number required?

19. Why should several secondary assignments be tried in sequential system synthesis, as well as several S-R or Tr, or combinations thereof?

20. Why can't a single x_1 input line enter both an S_1 and an R_1 of a flip-flop stage?

21. Obtain minimum terms for S_1, R_1, Tr_2, and z for the following secondary-assignment table:

$y_1 y_2$	x_1	x_2
00	01	00
01	01	11_{z_1}
11	01	00_{z_2}
10	----	----

22. Prepare a complete logical-circuit diagram for the terms obtained from Problem 21.

23. Prepare two other secondary-assignment tables from the one shown in Prob. 21 and obtain minimum terms of S_1, R_1, Tr_2, and z for each.

24. Prepare complete logical-circuit diagrams for the terms obtained from Prob. 23.

25. **a.** Did any of the new diagrams for Prob. 24 result in circuit reduction, or were more circuits involved?
    b. Which of the three diagrams had fewer circuits, and what logical switches were eliminated?

INDEX

227